Build Your Own Low-Cost Data Acquisition and Display Devices

Build Your Own Low-Cost Data Acquisition and Display Devices

Jeffrey Hirst Johnson

TAB Books
Division of McGraw-Hill, Inc.
Blue Ridge Summit, PA 17294-0850

FIRST EDITION
FIRST PRINTING

Library of Congress Cataloging-in-Publication Data

Johnson, Jeffrey Hirst.
Build your own low-cost data acquisition and display devices / by Jeffrey Hirst Johnson.
p. cm.
Includes index.
ISBN 0-8306-4349-4 ISBN 0-8306-4348-6 (pbk.)
1. Computer input-output equipment—Amateurs' manuals.
2. Information display systems—Design and construction—Amateurs' manuals. 3. IBM microcomputers—Equipment and supplies—Amateurs' manuals. 4. IBM-compatible computers—Equipment and supplies—Amateurs'. I. Title.
TK9969.J62 1993
006—dc20 93-10931
CIP

Acquisitions editor: Roland S. Phelps
Editorial team: Laura J. Bader, Editor
Susan Wahlman, Managing Editor
Joanne Slike, Executive Editor
Production team: Katherine G. Brown, Director
Patsy D. Harne,Layout
Sandy Hanson, Typesetting
Olive Harmon, Typesetting
Tina Sourbier, Typesetting
Cindi Bell, Proofreading
Design team: Jaclyn J. Boone, Designer
Brian Allison, Associate Designer
Cover design by Graphics Plus, Hanover, Pa.

EL2
4355

Contents

Introduction

If I had had this book some years ago I could have saved a good deal of time and effort. One reason for writing it was to put at your disposal the fruits of that labor. But that was not the only reason. There were two others that were at least as important. One was to fill a gap. Surprisingly, extended treatments of the standard input and output (I/O) ports on the IBM PC and compatible computers are few and far between. There are books that explain the serial port, but they do so from the standpoint of modems and telecommunications. Information on the standard serial and parallel ports as general-purpose input and output channels is scattered here and there and often treated in a superficial way. On top of that, the hardware is not without a few quirks, some of which one learns about only through experience.

The other motive, and in many respects the primary motive, was the conviction that the enormous potential of small computers for implementing modest data acquisition and control systems is largely untapped. The lumbering mainframe computer is already an endangered species. It was undone by much smaller inexpensive machines that made up in accessibility and economy what they lacked in power. Data acquisition systems are, I think, in somewhat of the same situation. The time is ripe for doing big things by means of thinking small.

The fact is that the PC family lends itself very nicely to interactions with the "real world." Once you begin to think of data input and output in forms other than just keyboards and printers, I think many applications will occur to you. At first I thought a little catalog of such applications might be helpful. But then I remembered the lesson of the microcomputer explosion of the last decade: that it was largely fueled by individuals and small groups thinking of things that hadn't been done and using small and inexpensive computers to do them. "Computing power for the people!" was a slogan one heard a lot in those early days. I decided it was better just to throw in suggestions and allusions in the course of the book.

Writing this book has been rewarding. Directly and indirectly it owes much to supportive colleagues at Duke University, a forbearing (and computer-savvy) wife, and an enthusiastic and remarkably patient editor. Without them this book would not now be in your hands.

❖1 Data acquisition

A computer isn't much use without data to work on. For most of us most of the time, data comes in from a keyboard or disk and goes out to a printer or plotter. Another common path is a modem and communication network. The common thread in these cases is that data is entered by human operators at some point, and the ultimate output is meant for humans.

Data acquisition, in contrast, has mainly to do with getting data to and from *processes*, not human operators. And it is more than just a matter of automating data entry. It is really about process interaction and control—everything from electronic test instruments to numerically controlled machinery and robotics. The field goes back to the early days of computers, but it has tended to remain a specialty. That is a pity, for it means that many ordinary users are unaware of the usefulness of data acquisition systems even for small-scale applications. The considerable promise of small computers in this regard has largely gone unrealized.

This book is about data acquisition and control using the IBM PC family, including true compatibles and the IBM PS/2 family. Because data acquisition necessarily involves data transmission, it is about that too. What sets this book apart from most other books on the subject is its emphasis on what can be done using only the standard I/O ports.

Scope and approach

Figure 1-1 stakes out the territory. It shows a very simple, but common, closed-loop control system—the temperature-controlled bath. Such systems are often implemented directly with analog hardware, but here a computer is included in the loop. The ability of computers to display and save data, analyze trends, and implement complicated temperature-versus-time protocols sets them far ahead of analog systems. The parts of the system are shown in the *data acquisition* box, and each will be discussed in detail. In practice, simple single-function control loops like the one in Fig. 1-1 are often just one element in a larger system. The ease with which groups of subsystems can be controlled together is another advantage of using computers.

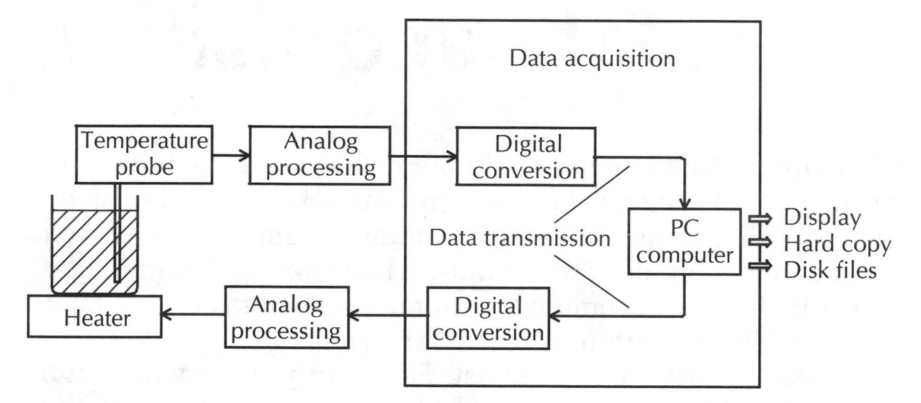

Fig. 1-1 *Example of data acquisition in a closed-loop control system.*

This book is concerned expressly and specifically with small systems. Not all data acquisition jobs are big jobs, and there is plenty of room at the bottom for quite modest systems. Modest size does not necessarily mean modest performance. In fact, just as the minicomputer and the PC have made the big mainframe machine something of an endangered species, so too are small data acquisition systems likely to become increasingly important.

Such an approach has many advantages. Economy is one. By using resources already built into the PC and by constructing some simple accessories, rather sophisticated systems can be set up at little expense. This is of considerable importance in these days of tight budgets. Another advantage is that it leads, I think, to a useful frame of mind. If setting up a data acquisition system means computers have to be opened up and expensive adapter cards installed, or complex cabling has to be run, or some sort of expert has to be brought into the picture, the tendency is to do it only for the "big" or "important" jobs. But modest setups using existing computers that can be hooked up quickly and easily, almost ad hoc, encourage the use of productive and efficient data acquisition techniques for the small jobs too. And these small jobs can often benefit just as much as the larger projects. Best of all, it encourages good old-fashioned tinkering, the kind of productive puttering that has been the basis for more than a few important advances.

Speaking of economy, one thing this book might help to do is to dispel certain myths. Too often the latest advance in PC hardware is billed as making older machines obsolete. The latest 486 machines are amazing, no doubt about it. But older machines—ATs, XTs, and even the original PCs—can perform quite satisfactorily in many data acquisition and I/O applications. There's a growing awareness that it doesn't necessarily make sense to throw out an older machine in good working condition and spend money on a newer one just to be able to run some fancy piece of software.

This book might also help the reader to put the matter of PC performance into perspective. A modern 386 or 486 machine executes many kinds of program code a great deal faster than the 8088 used in the original PC and can address vastly more memory. No doubt about it, they are impressive machines. But there is much less difference when it comes to I/O processing. For reasons that become apparent in the next chapter, specifying the "speed" of a computer is not a simple thing. In I/O-intensive operations an AT or 386 machine will not prove to be "blazingly" faster than a good 8- or 10-MHz XT. For purposes of data acquisition one should not reject out of hand an XT or even an old PC that happens to be available. The relevance of this to the matter of cost-effectiveness seems obvious.

This book tries to be useful to a wide range of readers. This is not just a matter of accommodating both the neophyte and the professional. Input and output is different from other aspects of computer technology because it is inherently a hybrid of hardware and software, and not everyone has experience in both. For this reason I have combined elementary introductions to the major topics with more detailed subsequent discussion.

In any case, this book gives you the tools to implement first-rate I/O functions in your programs and systems. All of the hardware designs and software routines were hammered out in the course of developing real-world applications, and megabytes of data have passed through them. Although there is plenty of practical information, it is purposely not presented in the form of complete stand-alone finished projects. (Some simplified completed designs and programs are included to show how everything fits together.) The aim throughout this book is to encourage experimentation and thinking. To give just one example of where this kind of creativity is needed, think about the enormous potential of inexpensive computer-driven systems for assisting the disabled.

Applications

It's hard to list the potential applications of data acquisition and transmission systems, not because there are so few, but because there are so many. A few application suggestions grouped into (loose) categories might be helpful.

The most obvious category is *data acquisition*—transforming some measurable quantity into digital form or digital information into some nondigital quantity. A digital voltmeter (DVM) with a computer interface is the simplest example. (Simple, but not trivial. Any quantity that can be converted to a voltage is fair game for measurement!) A program can plot the incoming data on the screen to produce an oscillographic display. By also storing it in memory, interesting sections can be replayed at will. Numerical processing (digital filtering, smoothing, statistical analysis, Fourier transforms, and so on) can also be applied. This kind of "digital oscilloscope" is so basic and so useful that it's used as the primary illustration for the hardware and software projects developed later in this book.

There's a lot to be said for thinking small. It's easy to overlook simple but useful applications. Consider the monitoring of important voltages and currents in a piece of equipment. Sometimes this is required—as in the station log of a radio transmitter —and in other cases it is helpful—continuous supervision lets fault conditions be caught immediately. Take the matter of monitoring the plate voltage and cathode current of the final stage of a radio transmitter. Figure 1-2 shows a simple setup to do that. Using a computer makes it trivial to keep track of the plate input power (just multiply the voltage and current), and makes it easy to implement automatic logging as well. The simple and inexpensive eight-bit (analog-to-digital) converters discussed in chapter 6 are ideal for such applications.

Many industrial and laboratory instruments contain some sort of computer interface. A case in point is one in which a commercial scanning densitometer was used to evaluate electrophoretic gels, in which different substances migrate at different rates along a lane or track and the optical density is proportional to their concentration. The basic output of the densitometer was accurate but rudimentary, so the serial interface in the

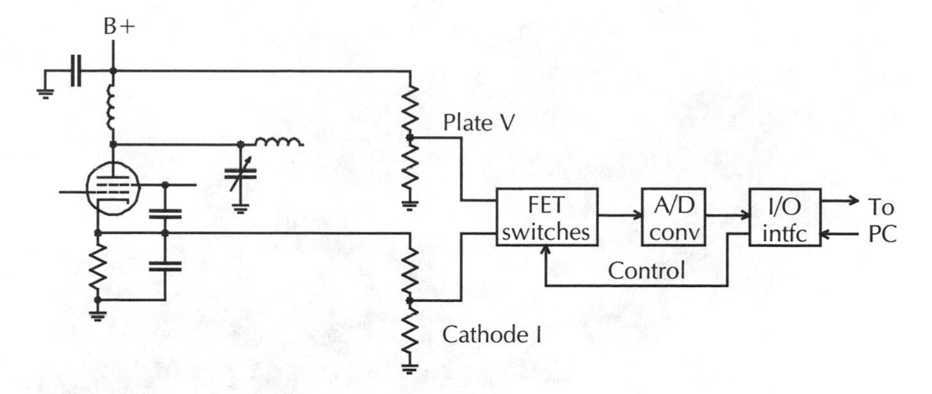

Fig. 1-2 Example of data acquisition for monitoring system parameters.

instrument was used to connect it to a PC running more sophisticated analytical software. This provided a larger color display with more information, as shown in Fig. 1-3. More importantly, it facilitated analysis, because the areas under parts of the curve (as in the hatched region in the figure) could be marked and evaluated.

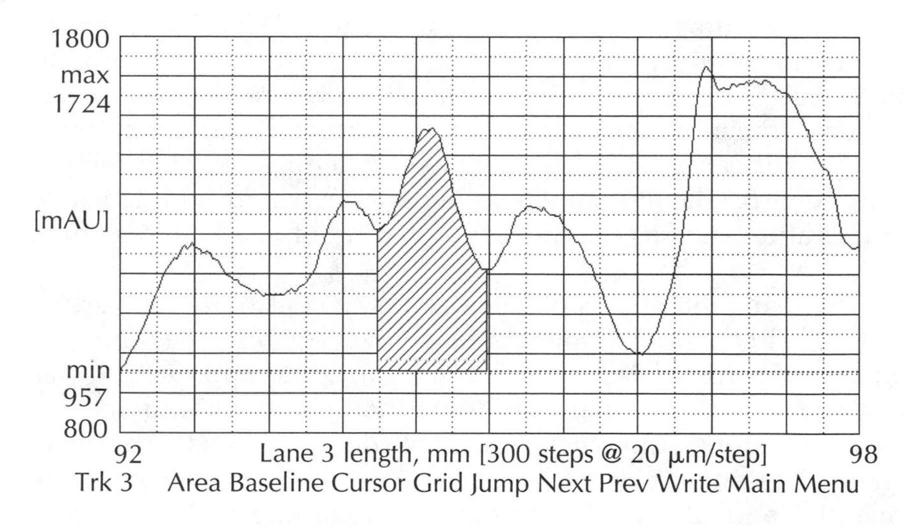

Fig. 1-3 Display of data obtained from an instrument through its computer interface.

Another benefit of customizing a system is the ability to add new functions and new ways of presenting data. In the application just mentioned, for example, the individual lane scans can be combined into a perspective presentation of an entire gel, as shown in Fig. 1-4, giving the user an immediate grasp of the global relationships among the lanes.

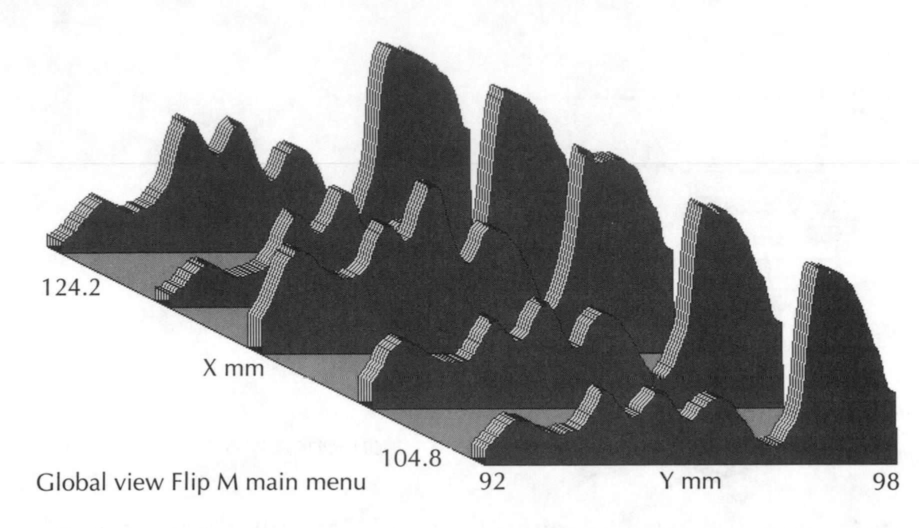

Fig. 1-4 *Data can be shown in many ways. The middle polygon is the same data as in Fig. 1-3.*

Control systems

An obvious area of application is in conjunction with the many modern laboratory and test instruments that provide a computer interface. But there are many other places where a computer can control something directly rather than through a human intemediary. This can be a simple thing like switching a light or relay—a system like the one shown in Fig. 1-1—or a more sophisticated application such as the numerical control of precision machine tools.

By using the results of calculations to control digital-to-analog (D/A) converters, you can take advantage of programmable analog control. Not only is it a more flexible approach, it has other benefits such as the ability to incorporate instrument correction factors. An extension of that idea is a digital waveform generator—in its simplest form, a D/A converter connected to the parallel port. By looping through a preset array of values and writing each one to the D/A converter, any arbitrary waveform can be generated.

An extension of this idea is the instrumentation system. Not so long ago the only way to make extensive measurements on a system was by hand. To measure the frequency response or distortion of an audio system, for instance, meant laboriously measuring point after point and writing the readings down on paper. Now these time-consuming tasks can be done automatically and more accurately and the results printed or graphed.

An example of a simple control system is the battery charger shown in Fig. 1-5. Battery chargers usually start off with a fairly hefty current to minimize charging time, then taper off to a very small "trickle" current as the battery reaches full charge. In the computerized system in Fig. 1-5, a D/A converter receives control data from the converter and provides an analog control signal to a power current regulator. A DVM with a computer I/O interface reports back the terminal voltage of the battery.

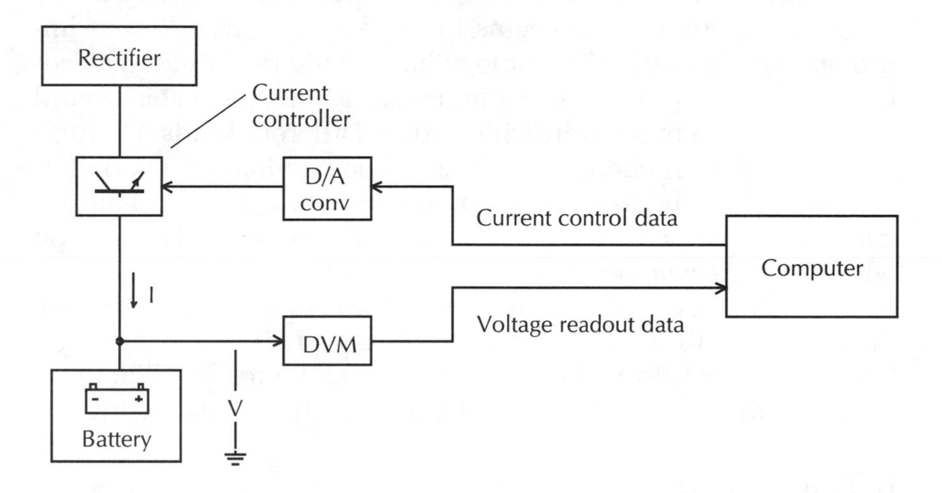

Fig. 1-5 *Example of a simple control system: a computerized battery charger.*

Using a computer for this purpose might strike you as a prime example of overkill. A good charging circuit is easily implemented with rather simple analog circuitry, and dozens of designs are available (it's a stock example in transistor and integrated circuit (IC) books). But not so fast. The computerized system is not without certain strengths of its own. The degree of charge is usually monitored by measuring the battery voltage. As the battery nears full charge, its voltage changes only a little. The high resolution of a digital voltmeter (such as the module described later in this book) allows very precise monitoring of battery voltage and therefore of the level of charge. Precise control of the charging current isn't important—within 10 percent is certainly good enough—but, even so, that capability is present. Moreover, the computerized system can do things that an analog system, even a sophisticated one, cannot. The software can be preset for various currents so that only a battery type need be entered by the user. Different rates of current versus battery voltage or time are simple to implement. A record of charging progress can be kept, which can identify batteries that are nearing the end of their useful life.

What really sets the computerized charger apart from an analog version is its capacity for various kinds of expansion. Several batteries can be handled at once by providing separate current control circuits for each one. Each D/A converter would have an addressable latch, and a field effect transistor (FET) multiplexer or a group of relays could connect the appropriate battery to the DVM. Computer control begins to make more sense. Recent improvements in battery technology have made rechargeable cordless power tools and the like practical, so such a system could have real value in an industrial setting. Different kinds of batteries with different charging rates could be handled easily. Batteries could even be assigned ID numbers and a simple database could keep track of their use and could provide statistics on which brands seem most reliable or most cost effective.

The real point of this example is not to present a specific project but to illustrate a method. When you are planning a system it can pay off to take a moment to ask yourself, "What else could it do?" and to play around a little with the possibilities.

I/O port drivers

An understanding of the standard I/O ports and how to use them fully gives you the knowledge to create your own high-performance I/O drivers that you can incorporate in the programs you write. For example, you could work up a serial port driver that provides the flow control (explained in chapter 3) that is essential for many peripherals, such as pen plotters.

Overview of data acquisition methods

There are many ways to set up data acquisition systems with varying degrees of complexity, performance, and (of course) cost. Sometimes the method is dictated by the application. For instance, getting input from a digitizing tablet that has an RS-232 output means using the serial port. When the situation is less clear-cut, choosing wisely depends on knowing what the options are and having a feel for the strengths and weaknesses of each. The following overview of some common approaches is offered to help provide some perspective.

The serial ports

The PC supports up to two standard serial ports, commonly known as the COM or RS-232 ports. Serial port cards are inexpensive, and most machines of recent vintage come with at least one serial port already built in.

Simplicity and economy are the hallmarks of data acquisition through the serial port. Cabling is inexpensive and noncritical. With a modem, data can be transmitted through the telephone network, making data acquisition as wide as the world. Considering these many advantages, it is easy to see why so many modern measurement and laboratory instruments and data subsystems support a serial interface. Incorporating a serial interface—a universal asynchronous receiver and transmitter (UART)—into equipment you build is not difficult, as chapter 6 shows.

The standard RS-232 interface provides bidirectional data transmission, and a pair of bidirectional control lines are also available. The serial port provides reliable data transfer at rates up to 3840 bytes per second (b/s), and some PCs can be pushed as high as 11,520 b/s.

The parallel ports

The PC supports up to three parallel ports. As with the serial ports, the parallel ports are inexpensive to add and quite often at least one is built in.

The standard parallel port is specifically intended as a printer interface, but it works well as a general-purpose parallel output. It is not always realized that the parallel port has a set of input lines also. Although the port is not designed for bidirectional operation (both input and output), chapter 5 discusses how it can be manipulated by software to provide good bidirectional operation. While this technique is not standard, it is very easy to implement and can be quite useful for setting up one's own systems.

Parallel data transmission is inherently faster than serial transmission (for a given bandwidth). The price that is paid is bulkier and more costly cabling and an inability to send data through a single channel, such as a telephone circuit. Depending somewhat on the type of PC, data transfer rates through the standard printer port of 5 kilobytes per second (kb/s) to more than 50 kb/s are possible.

The GPIB system

The general-purpose interface bus (GPIB), also known as IEEE-488, is very widely used for data acquisition, analysis, and automated test and control applications. It is specifically designed to support instrumentation systems, with up to fifteen interconnected instruments. Most modern instruments and data acquisition devices support the GPIB, at least as an option.

The GPIB is basically a parallel, bus-oriented bidirectional transmission system. Instrument control and data transmission functions are clearly distinguished, and an ingenious three-wire handshaking system controls data flow. The data transfer rate depends on each particular system. Although the GPIB is capable of more than 500 kb/s, the bus automatically runs at the speed of the slowest instrument connected to it.

For complex, high-performance instrumentation systems the GPIB is clearly the system of choice. Against these notable advantages must be weighed the fact that a GPIB adapter board, a relatively expensive item, must be installed in the PC. Devices on the bus must be connected using standardized cable assemblies. The software is also a little more complicated because the GPIB uses device addressing and various control commands in the course of its operation. When the advanced features of the GPIB are not required, the simpler techniques described in this book are worthy of serious consideration.

Data acquisition adapter cards

There is a very wide range of data acquisition and signal-processing boards that plug into the PC expansion slots. Many are actually complete data acquisition subsystems providing A/D and D/A conversion. Others provide some special function—for example, the video digitizer ("frame grabber") or the musical instrument digital interrace (MIDI) control system.

Special-purpose boards might be the best and sometimes are the only practical approach in specific applications. For more general cases, such as general-purpose A/D and D/A conversion, a disadvantage is that the board is restricted to one computer. To use a different computer, the machines must be opened and the board transferred. That makes quick ad hoc changes in systems more difficult. An alternative is to set up the signal processing parts of a system as independent units that can be connected to any available PC through the standard ports. A variety of hardware building blocks are presented later in this book that you can use to create such units, with the further advantage being that they can be tailored expressly to your needs.

Other methods

These data acquisition systems are the most common, but there are others that are appropriate for more specialized applications. Among these are integrating PCs with minicomputer or mainframe systems, interfacing with embedded control system, and various kinds of high-speed data links.

One case that deserves mention results from the increasingly common interconnection of several PCs on a local area network (LAN). Many LANs provide a means for low-level access to the transmission system, such as IPX/SPX in Novell NetWare. In this way the network can serve as a general-purpose data pathway. In installations where a LAN is already in place, a rather powerful yet inexpensive data system can be set up by combining LAN transmission with the data conversion and I/O methods described in this book.

Another system that has not hitherto been used much for data acquisition but which might be more important in the future is the Small Computer System Interface (SCSI). Originally used chiefly for block I/O to tape drives and hard disks, it has some interesting properties that make it potentially useful for general data I/O.

Plan of the book

By its very nature, data I/O brings together three areas of computer technology that otherwise are frequently dealt with more or less in isolation: (1) hardware, (2) software, and (3) communications and transmission. All of these subjects will be discussed in the course of this book. A survey of the territory covered seems in order.

The hardware comes first. Chapter 2 looks inside the PC and examines its architecture. Although I concentrate on the use of standard ports, some knowledge of the system bus and supporting subsystems is important. The interrupt mechanism is examined in particular detail. There is also a brief introduction to the 80*x*86 microprocessor for those who are not familiar with it.

Some key principles of data transmission are covered in chapter 3. I consider transmission as a system involving electrical signals, the way those signals are formatted, and the crucial matters of synchronization and flow control. I look at the two fundamental transmission methods: serial and parallel. The common RS-232 standard is the primary focus, though other forms of serial transmission are considered also. Parallel transmission concentrates on ordinary TTL-level (transistor-transistor logic) systems, including the Centronics printer interface specification.

After these preliminaries I get down to brass tacks and develop practical, working routines and set up real-word connections. Chapter 4 deals with the standard serial port, how to program it, and how to use it. It culminates in the presentation of the serial toolkit, a library of routines for control of the port and for fast, reliable I/O. Use of the toolkit is illustrated by a sample program that emulates a simple American National Standards Institute (ANSI) computer terminal, with further illustrations in the sample programs in later chapters.

Chapter 5 deals with the standard PC parallel port. Although the port is designed to be a printer interface, it can be used for general-purpose parallel output. It can also be used for input, and with the help of suitable software manipulations it can serve as a full eight-bit parallel interrupt-driven input. A library of routines is offered as the parallel toolkit.

It's no good having the ability to acquire data if there is no data to acquire. I turn again to hardware in chapter 6, this time in the form of simple building blocks that can be used to construct a variety of useful data acquisition units. Through the magic of modern integrated circuits, rather sophisticated devices can be put together by anyone who has a basic knowledge of electronics and can solder.

Chapter 7 is where I put all the pieces together as a system. First I am concerned with the mundane but crucial matters of interconnections, cabling, and the like. Then I develop a simple data acquisition program to illustrate how the toolkits and the conversion hardware can be combined into a useful system. Finally there are some suggestions for the creation of data acquisition and system control programs (a topic big enough for a book of its own). From that point on, it's the reader's turn: to rev up his or her imagination and get to work.

In the text I have generally followed the nomenclature and typographical practices current in the PC literature. For example, hexadecimal notation is indicated by an "h" at the end of the number, and binary by "b" (no suffix indicates decimal). Names of signals that are active low is problematic. The overbar convention makes a lot of sense, but is extra work in typesetting and possible only with contortions in word-processed documents. I have generally prefixed such names with "*"; for example, *STB. In some cases, such as the system bus names in chapter 2, the minus sign notation is so well established that I have kept it. I trust this minor inconsistency will not be too distracting.

A word about software code examples. They are in Pascal—specifically, Borland's Turbo Pascal. (When appropriate, cross-references to equivalent functions in the popular C and C++ languages are given.) Pascal was chosen for several reasons. One is that it is more nearly self-explanatory than most languages, making it easier to see just what is going on in the code. Another is that it is logically elegant and powerful. It contains more information (meta-information, really) than many other common programming languages. It is interesting that recent revisions of the venerable and popular FORTRAN and BASIC languages have become rather Pascal-like. Pascal is "strongly typed," which allows the compiler to catch many errors that other language compilers would miss. (Surely that qualifies as "power"!) Turbo Pascal version 4.0 or later is assumed. A few routines are in assembly language for superior performance or to take advantage of 80*x*86 commands not supported in most high-level languages, such as shift-through-carry. They are written in as generic a form as possible to allow the use of any standard 80*x*86 assembler. Alternative routines coded in Turbo Pascal are included for those who have no access to an assembler.

This book, then, is about tools. But it's up to you to apply them. Using your imagination is *your* responsibility. Don't get tied down to what you have seen computers do. Think about what you would like a computer to do, and then consider how you might actualize it. And don't disparage the trivial. Today's silly idea might be the seed from which tomorrow's brainstorm grows.

Under the hood

In this chapter you'll learn how data moves around inside the PC. It is a chapter mainly about mechanics. It includes a brief introduction to the 80x86 family of microprocessors. You'll take a look at the system ("expansion") bus, not from the point of view of designing adapter cards to work with it, but so far as it is relevant to using the I/O ports fully. The PC interrupt system is exceedingly important for I/O operations and it is explored in detail. Finally, the chapter takes a first look at some of the issues involved in moving data into and out of the PC.

Since the introduction of the IBM Personal Computer in late 1981, the PC has grown from a relatively simple but solid machine into a broad family of small computers offered by many vendors. Keeping the variations straight is of some importance. In this book the designation "PC" always refers to the whole family. When the original IBM Personal Computer is meant, it is called the "OPC." Unless indicated otherwise, "AT" indicates both the original 80286 AT and later versions using the 80386 and 80486 microprocessors.

The CPU

A clear understanding of basics is important. Let's begin with a look at how a computer works at a very elementary level. (Readers who are familiar with these matters might want to skip to the next section.)

A computer consists of four basic units:

- a *logic* unit, which performs the actual logical and arithmetic operations on data;
- a *data store*, which holds data and results;
- an *instruction store*, which holds a list of instructions; and
- a *control* unit, which processes instructions.

The instructions tell the logic unit two things: what to do and what data to do it to. Note that data is identified only by its location or *address* in the store. Instructions are likewise simply numbers, each of which activates the appropriate circuits in the control unit.

Suppose the computer is programmed to solve the equation $z = x + y$. Let's assume that x is at location 7, y is at location 3, and z is at location 24. The instructions might be, say, 1005 3 1004 7 1012 1006 24, which could mean "add the value stored at #3 to the value stored at #7 and put the result in #24." The instructions, or *code*, are stored sequentially. The control unit gets one, does what is indicated, increments its code address value, and fetches the next instruction.

These functions are usually grouped together and called the *central processing unit* (CPU). CPUs have a small and very fast internal store called *registers*. Some are for temporary storage and intermediate results—data is moved into one or more registers from the main store (i.e., memory), manipulated, and then put back in memory. Other registers are used to hold indices to memory locations. There is a register that holds the address of the current instruction and a register that holds certain processor status information ("flags").

Because the data are numbers and the instructions are numbers, many (though not all) computers, including PCs, store them together in memory. How does the CPU tell whether a number is a data item or an instruction? It can't. It is up to the program to ensure that the instruction register is loaded with a valid code address.

Ordinarily instructions are executed in sequence, one after the other. But often we would like to do different things depending on the result of a calculation—the if . . . then construct. This is called *branching* and is done by changing the address in the instruction register to the address of the target instruction. Fortunately the details, which can be messy, are handled automatically by language compilers.

A very important special case is the *subroutine*. These are sections of code that are used over and over (boilerplate, as it were). When a subroutine is called, the current instruction address is saved in memory before branching to the subroutine. When the subroutine is done, it "returns" and the saved instruction address is restored and the main program takes up where it left off. This is a very powerful method. Programs in a number of modern languages, such as Pascal and C, consist entirely of subroutines (procedures and functions).

In many computers, including the PC, the CPU sets aside an area of memory, called the *stack*, to keep track of subroutine parameters and return addresses. Items are saved by "pushing" them on the stack and retrieved by "popping" them off. Each "pop" gets whatever was most recently "pushed." This makes the stack extremely useful for temporarily saving the contents of CPU registers. You will see later that this is vital to certain powerful I/O techniques.

The 80*x*86 microprocessors

Intel introduced one of the first sixteen-bit microprocessors, the 8086, in 1978. (Morse [1982] is an excellent introduction to the 8086.) The original PC (OPC) used the Intel 8088, identical to the 8086 except that it uses an eight-bit data bus. (Sixteen-bit items are transmitted as two eight-bit slices, one after the other.) The IBM PC XT and the low-end models of the IBM PS/2 use the 8088 also.

The IBM PC AT was the first model to use a different processor, the 80286. Like the 8088 it is a sixteen-bit processor, but it expanded the address capacity from the twenty bits (1,048,576 addresses) of the 8088 to twenty-four bits (16,777,216 addresses). In its "real" mode it is essentially a bigger and faster 8086. It can also run in a "protected" mode in which memory references are indirect, allowing memory to be divided into segments and assigned different access rights. In this mode several programs can run together without the risk of overwriting each other.

The Intel 80386 marked a major change in this family. It is a full thirty-two-bit device, and it can directly address some 4,294,836,224 locations in memory. In real mode it is very much like the 8086 and 80286, and generally similar to the 80286 in protected mode. It also has a "virtual" mode in which it is, in effect, an array of 8086 processors. The 80486 is similar to the 80386 but incorporates a floating-point unit.

The degree of compatibility among these four very different microprocessors is remarkable. For that reason, most everyone refers to them as a group simply as the "80*x*86" family.

Real and protected mode

Heretofore PCs with 80286 or later processors were nearly always run in real mode and ran standard DOS. As such they were essentially larger and faster versions of the original PC. Recently there has been an increasing use of protected mode operation, either under one of the "DOS extenders" or with operating systems such as Microsoft Windows or IBM OS/2. Because the protected mode allows programs to be insulated from one another, it is particularly advantageous for running several programs at once (multitasking). You should be aware, however, that in several important respects a PC running in protected mode is a different machine. One very significant difference is that an application program cannot directly address memory. The function of the CPU segment registers changes. They become *selectors*, indices into a table of memory assignments. Systems that provide a "DOS window" or "compatibility window" set up special selector entries for some common memory addresses such as the video screen buffer.

Which is better, real or protected mode? There is no right answer to that question. They are simply different, and neither is better than the other. Real mode allows a much more direct and intimate connection between software and hardware. The result is a tightly integrated package that, if properly done, can yield a level of performance second to none. Programs can be simpler and more compact. Real-mode operating systems, such as standard DOS, are much smaller and thus leave more disk and memory space free for application programs (no matter how big the disk and no matter how much memory, more of it will be available). From the practical user's standpoint, the result is something like this: protected mode can do many things at once and do them pretty well; real mode can do only one thing at a time, but can do it superbly.

All of the material in this book assumes a PC running in real mode.

The 80*x*86 registers

Table 2-1 shows the common register set in the 80*x*86 family. Looking at the 80*x*86 at the register level might seem a little far afield in a book devoted to I/O, but the CPU registers will be used in software routines. The four registers A, B, C, and D are general purpose; they can hold any kind of variable. Each can be used either as separate eight-bit registers (e.g., AL, AH) or as a single sixteen-bit register (e.g., AX). They also have a few dedicated uses. For example, port I/O goes to and from the A register, and the CX register is used to hold the number of times a loop is to repeat.

Table 2-1 80*x*86 registers.

16-bit	8-bit	Purpose
		General and special-purpose
AX	AH, AL	Arithmetic results; string and port I/O
DX	DH, DL	Port number; longints
CX	CH, CL	Loop counter
BX	BH, BL	Address indices
		Addressing
BP		Address indices in stack
SI		Source for string instructions
DI		Destination for string instructions
SP		Stack pointer
DS		Current data segment
ES		General segment
SS		Current stack segment
		Instructions and CPU control
CS		Current code segment
IP		Instruction pointer
Flags		Processor status

The addressing registers are used to access memory locations. The BX, SI, and DI registers are used to index into arrays or strings. (They can also be used as additional general-purpose registers.) The CS, DS, ES, and SS registers hold the upper sixteen bits of addresses. Although Intel refers to them as "segment" registers, this is misleading. True segmented memory slices the address range into specific areas that might incorporate protection to control or prevent access to those areas by less "privileged" code. The 8088 has "flat" addressing, not segmented (the 80386 and 80486 in protected mode do). Now a sixteen-bit register by itself can address only 2^{16} (64K) locations. To get a greater range, such as the twenty bits of the 8086 or the twenty-four bits of the 80286, two registers have to be hooked together—like a long (double precision) integer. Instead of putting the lower sixteen bits of an address in one register and the remaining bits in the segment register, Intel chose to overlap them. This allows addresses within a 64K range to be accessed with only one register.

The importance of the segment registers for our purposes comes in interrupt routines. To address data from within interrupt code, both the offset and the segment registers have to be set. A complication is that the segment value depends on where in memory DOS has placed the program. In chapter 4 you will see how this is done.

The CS (segment) and IP (offset) registers contain the address of the current instruction. The flags register holds the processor status and control flags (each flag value is a 0 or 1 of a specified bit).

Computer busses

A computer isn't of much help unless it can interact with the external world through various input and output devices. These I/O interfaces and the equipment connected to them are referred to as *peripherals*. (Memory is a peripheral, strictly speaking, but it is so fundamental to a computer that it is normally associated with the CPU.) Because there are usually several peripherals, a way has to be provided for them to share the CPU. This is done by connecting all of them to a kind of party line called a *bus* (from the Latin *omnibus*, "for all").

Figure 2-1 shows a system bus in its most general form. Note that the system bus is composed of three smaller busses, one each for data, addresses, and control. Each device is designed so that it remains disconnected from the data bus except when it is activated (this is how many devices can share a bus). The selection is accomplished by assigning an address to each device. Each device monitors the address bus. The control bus contains several lines that control data flow—from device to CPU versus from CPU to device, and so on.

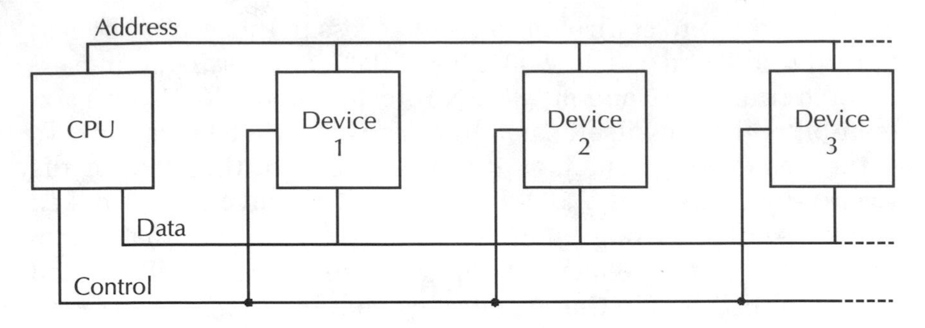

Fig. 2-1 Skeleton of a computer bus.

The innards of a typical peripheral device are shown in Fig. 2-2. An address decoder, an array of logic gates, responds when the address assigned to the device is present and tells the interface buffer (transceiver) to connect to the data bus. A line in the control bus tells the device whether this is a write (CPU to device) or a read (device to CPU) operation, and sets the direction of transmission through the transceiver appropriately.

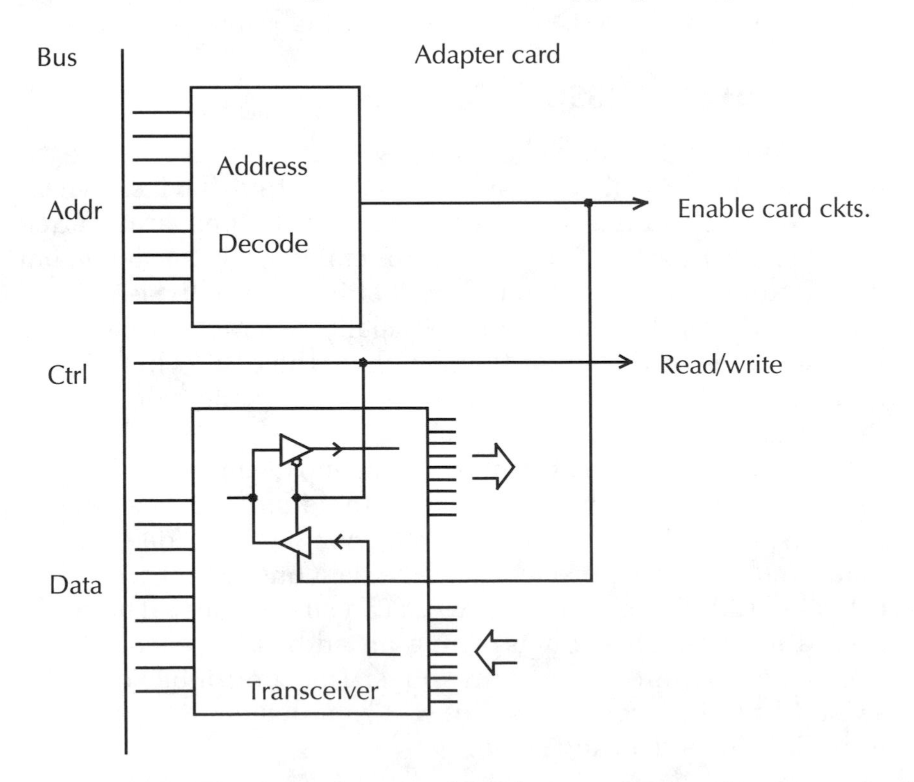

Fig. 2-2 Basic elements of the device-to-bus interface.

The 80x86 has two sets of data-moving instructions: a general-purpose primary set (the memory set) and a bare-bones secondary set (the I/O set). This allows two different busses to be set up. In some systems, including the PC, they are superimposed. For this reason the I/O bus is better designated as the *I/O channel*. One advantage of this arrangement is that the busses can operate at different speeds. The main (or memory) channel runs at the full CPU speed, but the I/O channel can run more slowly to accommodate sluggish I/O devices. The control bus consequently has two sets of read and write control lines, one pair for memory devices and one pair for I/O devices.

Peripherals that use the memory bus are said to be *memory-mapped*, and those that use the I/O bus are *I/O-mapped*. A peripheral can use both, as do the video cards in the PC.

The PC system bus

Although the focus of this book is on input and output using the built-in ports, a general knowledge of how the PC system busses work and of I/O address assignments ("what's where") is helpful. It should be of help in tracking down and resolving the signal conflicts that sometimes occur when adding plug-in adapter boards to a PC. By the way, what is here called the "system" bus is what IBM's documents call the "expansion" bus.

One of the main reasons the original PC was an instant hit was undoubtedly the fact that the system bus is readily accessible and its operation is documented. The original eight-bit bus was rather simple, not much more than an extension of the CPU bus together with some supporting functions such as interrupt and DMA lines. The system ("expansion") bus is brought out to several sixty-two-pin connectors into which adapter cards can be plugged. This bus was carried over into the XT and the lowest models of the IBM PS/2.

The first major change in PC architecture came with the AT, which used the 80286. The data bus was expanded to sixteen bits and the address bus to twenty-four bits, but to maintain compatibility—very wisely, in retrospect—the additional bus lines appeared on a second connector. This arrangement was widely adopted by makers of compatibles and has become known as the industry standard architecture (ISA).

When IBM introduced the PS/2 series, the upper-end models used a new bus—the Micro Channel architecture (MCA). It is considerably different from the ISA bus (among other things, it is more or less asynchronous). IBM has retained this proprietary design for its own use. Several of the major PC-compatible makers then introduced another new bus, dubbed the Extended Industry Standard Architecture (EISA). Like the IBM MCA bus, it can achieve substantially faster speeds than the plain ISA bus. Remarkably, a computer using the EISA bus remains compatible with the ISA design. Even the adapter board connectors are compatible; the extra lines are added between the connector fingers. (For a summary of the EISA see Glass [1989].)

ISA system bus signals

The following paragraphs summarize the signals in the eight- and sixteen-bit system busses. The names generally follow the IBM technical documentation. Electrically, the signals are standard 74LS TTL levels (low, 0.8 volt maximum; high, 2.0 volt minimum). A minus sign indicates that the signal is active low. Signal flow is designated by [O] for output from the bus (i.e., into an adapter), [I] for input, and [I/O] for bidirectional.

- OSC—[O], 14.31818-MHz clock.
- CLK—[O], system clock. Note that this signal has a 33 percent duty cycle. In the IBM OPC, XT, and AT it is simply the CPU clock. In 386 systems it is usually an independent clock, typically between 6 and 10 MHz. Some XT and AT compatibles also use an independent clock.
- RESET DRV—[O], reset signal for adapter boards.
- A0–A19—[O], address lines. A0 is LSB. Controlled either by the CPU or the DMA controller.
- D0–D7—[I/O], data lines 0 to 7. D0 is LSB. These lines are bidirectional and must be driven by three-state devices.
- ALE—[O], address latch enable. Used to latch valid addresses. Adapters can use it together with address enable (AEN) to indicate a valid processor address.
- IOCHK—[I], I/O channel check. Used by adapters to force a nonmaskable interrupt (NMI) in the case of a fatal and unrecoverable error. The basic input/output system (BIOS) will display it as a memory parity error. It must be driven by an open-collector or three-state driver and is pulled low to force an NMI.
- IOCHRDY—[I], I/O channel ready. Pulling this line low inserts I/O wait states, each of which is the period of the system clock. It must be driven by an open-collector or

three-state driver. If used, it should be driven low as soon as the card detects its valid address and a read or write command. It should not be kept low for more than ten system clocks.

- IRQ2–IRQ7—[I], interrupt request 2 to 7. The line is raised high to initiate a hardware interrupt. It must remain high until it is acknowledged by the CPU. Adapter cards cannot share the same IRQ line. (On the AT, IRQ2 is actually request line 9 on the slave Programmable Interrupt Controller (PIC). The BIOS revectors the interrupt request to imitate an IRQ2.)
- –IOR—[O], I/O read. When this line goes low, the adapter should read data from the data bus.
- –OW—[O], I/O write. When this line goes low, the adapter should drive its data onto the data bus.
- –SMEMR—[O], memory read. When this line goes low, a memory or memory-mapped card should read data from the data bus. On the AT it is active only for addresses less than 1 Mb.
- –SMEMW—[O], memory write. When this line goes low, a memory or memory-mapped card should drive its data onto the data bus. On the AT it is active only for addresses less than 1 Mb.
- DRQ1–DRQ3—[I], DMA request 1 to 3. The line is raised high to initiate a direct memory access (DMA) transfer cycle. It must remain high until it is acknowledged by the corresponding DMA acknowledge (–DACK) line. There is no line for DMA channel 0, which is dedicated to memory refresh in the OPC and XT.
- –DACK1–DACK3—[O], DMA acknowledge. These lines acknowledge a valid DMA request from the corresponding request line. DMA channel 0 is used for memory refresh on the OPC and XT. Refresh cycles are controlled by separate hardware on the AT.
- –REF—[O], refresh. This is used to initiate refresh cycles of dynamic memories. (On the OPC and XT it is actually –DACK0.)
- AEN—[O], address enable. A low level indicates that the CPU is in control of the system bus. When high, the DMA controller controls the bus.
- T/C—[O], terminal count. This line is pulsed high by the DMA controller at the end of a DMA transfer to indicate that the transfer is completed.

- –CARDSELECTD—[I], card selected. This line was used in early IBM PCs and XTs by system expansion cards in motherboard slot J8. In the IBM AT this pin (B8) was redefined (see below under 0WS) and bussed to all card connectors. XT compatibles might ignore this line or use it in the same manner as in the AT.

The sixteen-bit bus extensions added in the AT and located on the supplementary card connector are as follows:

- SBHE—[O], system bus high enable. A high level indicates that the upper eight data lines (D8–D15) are active. It is used for sixteen-bit data transfers.
- –MEMR—[O], memory read. This signal is the same as –SMEMR on the eight-bit connector except that it is active for all addresses including those above 1 Mb.
- –MEMW—[O], memory write. This signal is the same as –SMEMW on the eight-bit connector except that it is active for all addresses including those above 1 Mb.
- DRQ0,DRQ5–DRQ7—[I], DMA request 0 and 5 to 7. The line is raised high to initiate a direct memory (DMA) transfer cycle. It must remain high until it is acknowledged by the corresponding -DACK line. These are similar to the DRQ signals on the eight-bit connector except that they perform sixteen-bit DMA transfers.
- –DACK5–DACK7—[O], DMA acknowledge. These lines perform the same function as the –DACK lines on the eight-bit connector.
- –MEMCS16—[I], memory chip select 16. A card can pull this line low (an open-collector or three-state driver must be used) to select a sixteen-bit memory-mapped data transfer on the current bus cycle.
- –IOCS16—[I], memory chip select 16. A card can pull this line low (an open-collector or three-state driver must be used) to select a sixteen-bit I/O data transfer on the current bus cycle.
- –MASTER—[I], bus master. This line can be forced low after a DACK5 to DACK7 signal to allow a card to get control of the system bus. The card must obey all timing requirements of the bus specification and must relinquish the bus after 15 μs at most.
- IRQ10–IRQ12, IRQ14–IRQ15—[I], interrupt request 10 to 12 and 14 to 15. These lines are similar to the IRQ lines on

the eight-bit connector, except that they are connected to the slave PIC.

- LA17–LA23—[O], unlatched address 17 to 23. Unlike address lines A0–A23, these lines are active only during the CPU ALE. They can be used to reduce delay in address decoding in certain special cases.
- D8–D15—[I/O], data lines 8 through 15. These lines are the upper eight bits for sixteen-bit data transfers. They are bidirectional and must be driven by three-state devices.

The PC I/O channel

The memory and I/O channels of the PC system bus physically overlap and share the same address and data lines. They are distinguished from one another primarily by different sets of control signals, the memory strobes (–MEMR, –MEMW) versus the I/O strobes (–IOR,–IOW).

The speed with which data can move over the system bus is a function of the cycle time for a read or write operation. In the 8088, a transfer between the CPU and memory or I/O ordinarily takes four clocks. In the OPC with its 4.77-MHz clock the total cycle time is 840 ns. In the OPC and XT, however, logic on the motherboard automatically inserts an additional clock period—a wait state—giving an I/O cycle time of five clocks (1.05 μs with a 4.77-MHz clock). This means that I/O transfers are a little slower than memory transfers. In AT and faster XT clones the situation is less clear. In general, most such machines keep the I/O transfer rate in the same general neighborhood. Some designs insert additional wait states on I/O cycles; others use a separate bus clock. (These I/O wait states should not be confused with the memory wait states some faster machines use to allow for slower RAM.)

Although bus timing is of concern primarily in designing adapter cards, it does have a substantial effect on the overall I/O performance. In general, industry practice seems to be to keep the I/O cycle time around 1 μs. This sets a ceiling on data transfer rates. Every I/O operation is a combination of port access, memory access, and software routines. Using a higher clock rate or a more efficient processor (such as the 80286, 80386, and 80486) shortens memory access and execution time, sometimes dramatically. But if the I/O cycle time is not much different, a kind of diminishing return sets in. The total time can never be less than the I/O time, no matter how fast the processor is.

The bottom line is that replacing a computer with a faster one might not give you the increase in I/O speed you might expect. Published figures on how "fast" a computer is are almost always based on the execution speed of standard software routines. The improvement in input and output that you will get depends on the relative proportion of I/O time and execution time, which in turn depends on the specific software routines controlling the I/O. This effect is shown in Fig. 2-6 later in this chapter.

I/O address map

The principal I/O addresses are shown in Table 2-2. Some of the widely used addresses, such as the COM1 and COM2 serial ports, have become de facto standards, but unfortunately there is much less uniformity among more specialized adapter boards. Many boards have jumpers that can be changed to select different I/O addresses. Table 2-2 should help in tracking down and resolving conflicts.

Table 2-2 I/O address map.

Address (hexadecimal)	Device
	System addresses (decoded in blocks of 32)
000–00F	DMA controller (slave in AT)
020–021	PIC (XT); master PIC (AT)
040–043	Timer
060–063	PIO (XT); keyboard (AT)
070–071	CMOS RAM, part of real-time clock (AT only)
081–083	DMA page registers, channels 1–3
087–08F	DMA page registers, channels 0, 5–7 (AT only)
0A0	NMI mask register (XT); slave PIC (AT)
0A1	Slave PIC (AT only)
0C0–0DF	DMA controller (AT only)
0E0–0EF	Reserved
0F0–0FF	Math coprocessor
	I/O channel addresses (AT only)
100–16F	Reserved
170–177	Fixed disk 1 (AT only)
1F0–1F7	Fixed disk 0 (AT only)
1F9–1FF	Reserved

Table 2-2 Continued.

	I/O channel addresses (general)
200–20F	Game card
210–217	Expansion card (OPC, XT only)
220–22F	Reserved, or COM expansion
230–24F	Reserved
278–27F	LPT2 printer
2E1–2E3	Reserved, or data acquisition 0
2F0–2F7	Reserved
2F8–2FF	COM2 serial
300–31F	Prototype board (custom user devices)
320–32F	Fixed disk
360–36F	Network adapter
378–37F	LPT1 printer
380–3AF	Synchronous serial (SDLC or BiSync)
3B4–3BA	Monochrome video (MDA) control
3BC–3BE	Monochrome card printer port
3BF	Hercules configuration register
3C0–3D3	EGA, VGA control
3D4–3DF	CGA, EGA, VGA control
3E0–3E7	Reserved
3F0–3F7	Floppy controller
3F8–3FF	COM1 serial
	"Extended" addresses
3220–3227	COM3 serial
3228–322F	COM4 serial
42E1–42E3	Data acquisition 1
62E1–62E3	Data acquisition 2
82E1–82E3	Data acquisition 3
A2E1–A2E3	Data acquisition 4
E2E1–E2E3	Data acquisition 5

The 80*x*86 CPUs can address 64K I/O locations, but the implementation of the I/O channel in the PC family is more limited. Although 64K can indeed be addressed in the PC, the standard cards originated by IBM decode only the first ten address lines, giving just 1024 unique addresses. This means that standard adapter cards will respond to any address greater than 1023 (3FFh) whose first ten bits happen to match the ten-bit card address. A further limitation is that the lowest I/O addresses are used to control a number of system functions on the motherboard. On the OPC and XT, the boundary is at 200h; on the AT, it is at 100h. This leaves some 512 addresses available for adapter

cards to use (768 on the AT). The temptation to "steal" some unused addresses from the reserved system area should be resisted.

It is possible to use I/O addresses above 3FFh provided they are chosen with care. The value of the lower ten bits must fall at an unused address in the legal I/O channel range of 200h to 3FFh (100h to 3FFh on the AT). Some commercial adapter cards seem to use this ploy, and they are listed in Table 2-2 under "extended addresses." The possibility of unexpected side effects from address conflicts should be borne in mind.

Interrupts

Unlike files, where many bytes of data are transferred in one operation by a single write or read command, port I/O is generally handled byte by byte. Sending output from a program to a peripheral device is straightforward. When the program has data ready to send to the screen, printer, or whatever, it simply sends it. (If the device is busy, the program can wait and try again.) Getting input is another matter. Data can come, in general, at any moment. A program must be able to fetch it before the next byte of data arrives or it will be lost.

One way to do this is with a loop that constantly checks the status of each input port being used. If nothing is coming in at the moment, the next port is checked. If there is data ready to be processed, it is fetched from the port and the appropriate subroutines to handle it are called. This method is known as *polling*. It is simple, and a good choice in some cases, but it has its limits. All input-processing subroutines must be fast enough to allow every port to be polled frequently enough to avoid losing data. For this reason, polling is usually out of the question if the received data is being written to a file. The DOS subroutine that writes to disk can take as many as three or four seconds in the case of a floppy. The method is inherently sequential: the next port cannot be checked until the processing of the current one is finished.

The polling method gets input only when the program is able to ask for it. The program is the one in charge. What is needed is for the input to be in control. This can be done by an *interrupt*, which is a signal initiated by an input device that tells the CPU to stop what it is doing, jump to a special input subroutine that processes the input, and when that has been done to take up where it left off. It is similar to the way we handle telephone calls. The phone rings when somebody wants to talk to us. We stop what we are doing, talk, and then return to our task. If phones didn't ring you would have to use polling to get your calls, picking up the handset every few seconds and listening to find out if anyone was there.

Polling and interrupts are not exclusive alternatives. On the contrary, they work splendidly together. A subroutine activated by an interrupt can put each incoming byte in an array in memory (a *buffer*) as it comes in. The program can then use polling to check the buffer and fetch data from it. (This is exactly how reading a keyboard character works. The BIOS maintains a small buffer that is loaded by the keyboard interrupt each time a key is pressed.) As long as the buffer is large enough to hold all of the incoming data between polls, nothing is lost.

Consider again the matter of receiving a stream of data and writing it to disk. Suppose, for example, that you need to receive serial data at 9600 baud (typically 960 b/s). Allowing three seconds for a floppy disk access means holding some 3840 bytes, so you would set up a buffer of, say, 4K. Incoming bytes would be put immediately into the buffer by an interrupt routine. You would fetch bytes to be written to disk by polling not the input device, but the buffer.

How interrupts work

Interrupts have acquired something of a reputation as a rather arcane topic. It is true that there are certain constraints that have to be observed in order for an interrupt routine to work properly. But they are really not difficult to use, and use well. The key to success is a solid foundation of understanding of the principles on which they rest. This chapter looks at the hardware side. The software aspects are examined later in chapter 4.

The way the CPU handles interrupts is much like the way ordinary subroutines are handled, although there are some important differences. Subroutines called by interrupts are often called *interrupt service routines* (ISRs).

In high-level programming languages (and in assembly language, for that matter) most of the details of activating a subroutine are handled behind the scenes. In many languages a subroutine is activated by specifying its name; in others a line number or similar reference is used. The compiler then issues the appropriate machine-language instructions: in essence, a CALL instruction and the subroutine address (the location in memory of its first instruction). When the CPU gets a CALL instruction, it saves the current code address by pushing CS and IP on the stack and then jumps to the subroutine by setting CS and IP to the subroutine address. The last instruction in a subroutine is a RETurn. This reverses what was done in the CALL. The CPU pops CS and IP, fetches the next instruction, and resumes.

Please excuse a brief but important digression. The form of call just described uses the full address (up to thirty-two bits) of the subroutine and is known as a "far" call. It can be any address of which the CPU is capable. The 80*x*86 also provides a "near" version that uses only a sixteen-bit offset in the current code segment. This limits the address range to a 64K neighborhood, so the subroutine must be "nearby." Because the value in CS will not change, only IP is pushed. Note that all interrupts are far calls, meaning that any ISRs you write *must* be written as far procedures. This is pursued further in chapter 4.

In the case of an interrupt, a signal on the INT input pin of the CPU takes the place of the CALL instruction. But how does the CPU find out the address to jump to? The 80*x*86 finds out by asking the device that generated the interrupt. The interrupting device is expected to respond by putting a value on the data bus that tells the CPU where to jump to begin the ISR. In many processors, including the 80*x*86s, this value is not the actual address of the ISR, but rather an index into an array of addresses. This index number is commonly called the *interrupt number* or *type*. This indirect method of getting the ISR address is known as a *vectored interrupt* and the addresses in the array of addresses are called *interrupt vectors*. A vectored interrupt system has several advantages that outweigh its greater complexity. The 80*x*86 provides for 256 interrupt vectors, so a single byte suffices for the interrupt number as against the four bytes needed to hold a complete address. Even more important is that programs can change the interrupt vector to point to a different ISR. Programs can even chain or link ISRs together; for example, "pop-up" or "hot-key" utilities like Borland's Sidekick insert themselves into the keyboard interrupt. This is one of the things that makes the PC such a versatile machine.

The 80*x*86 has another trick up its silicon sleeve. It allows interrupt routines to be activated by software commands. The instruction INT, followed by an interrupt number, calls the corresponding ISR just as if the interrupt had originated in hardware. Software interrupts are how application programs communicate with the ROM BIOS and DOS, and most programs make extensive use of this powerful system of software interrupts. Many program language compilers for the PC offer a way to use software interrupts, such as the **Intr()** procedure in Turbo Pascal and the **int86()** function in Turbo C. Chapters 4 and 5 explore software interrupts in more detail.

Once the CPU has obtained the address of the interrupt subroutine from the interrupt vector table, it first pushes the flags register on the stack. It then proceeds as it would with a normal subroutine (far) call, pushing CS and IP and jumping to the ISR. When the ISR has finished, its last instruction is not an ordinary return but an IRET (interrupt return). This tells the CPU to pop the flags also. Another difference is that the CPU turns off interrupts by clearing the interrupt flag (after the existing flags have been saved). The ISR can enable other interrupts by setting the interrupt flag (IF) with the STI instruction. No STI is needed after an IRET because the original flags are restored by the IRET, and had the interrupt flag not been set, the interrupt would never have occurred.

A program can turn all interrupts off and on by using the CLI and STI commands. The CLI command clears the CPU interrupt flag, and the CPU then ignores all interrupt requests. The STI command sets the flag, enabling interrupts. Interrupts are sometimes turned off (briefly, please!) in critical sections of code such as very tight, fast loops. Few program languages support these instructions directly. Many do, however, allow some form of "inline" code that allows machine language (usually as hexadecimal numbers) to be embedded. The CLI is 0FA hex, and the STI is 0FB hex. In Turbo Pascal, for example, CLI would be `inline( $FA );` and STI would be `inline( $FB );`.

In addition to the normal hardware interrupts just discussed, the 80*x*86 provides a special interrupt that cannot be turned off—hence its name, the nonmaskable interrupt (NMI). The 80*x*86 calls interrupt 2 in response to an NMI. Its usual use is for emergencies, a kind of panic button for the system. In the PC family it is triggered by memory parity failures, and in the XT by activation of the –IOCHK line on the bus.

The interrupt controller

Because there are several devices in a PC that generate interrupt requests, the interrupt input of the CPU cannot be used directly. An IC known as an *interrupt controller* is placed between the CPU interrupt input and the various devices that generate interrupt requests. Like a receptionist in an office, it handles incoming requests and can rank them in order of importance or ignore some of them altogether. It also holds the interrupt number assigned to each of its inputs and provides that number to the CPU when activating an interrupt. The controller used in all versions of the PC and in the IBM PS/2 is an Intel 8259A programmable interrupt controller (PIC).

The 8259A is a sophisticated device with many modes of operation. Programming the PIC is a rather complicated, not to say convoluted, process—especially because Intel chose to provide just one address line for accessing its several internal registers. Fortunately most of these details are of no concern, and only the setup used in the PC is described. Indeed, its operating modes should not be changed from the ones set up by the ROM BIOS because the PIC services certain system devices as well as general I/O ports. Those who seek the gory details that are bypassed here are referred to the 8259A data sheet.

It is enough to note that the PIC in the OPC, XT, and AT is initialized to operate in edge-triggered mode, uses standard priority, and requires an end-of-interrupt command. However, in IBM PS/2 computers using the IBM MCA, level-sensitive IRQ lines are used.

The major functions of the 8259A are shown in Fig. 2-3. Each of the eight interrupt request lines is connected to a latch, forming the interrupt request register (IRR). Raising a given IRQ line from low to high sets the corresponding latch. The interrupt mask register holds a byte that enables or disables individual IRQ inputs. Each bit of the mask controls an input: bit 0 (the lsb) controls IRQ,

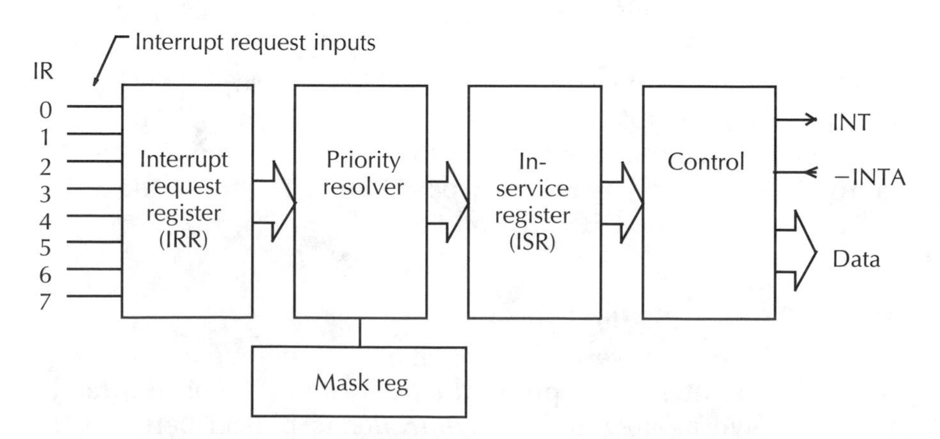

Fig. 2-3 *Anatomy of the 8259A programmable interrupt controller.*

bit 1 controls IRQ1, and so on. If a bit is clear (0), the corresponding IRQ line is enabled and can generate an interrupt; if it is set (1), that IRQ is disabled and has no effect.

The priority resolver selects which of several simultaneous interrupt requests will be passed on to the CPU. IRQ0 has the highest priority and IRQ7 the lowest. Pending interrupts are serviced in order, from highest priority to lowest. Another set of latches, the in-service register, holds the current status of each interrupt channel. If an interrupt request is approved by the priority resolver, the corresponding latch in the ISR is set. Finally an interrupt request is issued to the CPU on its INT pin. After the ISR has finished, an end-of-interrupt (EOI) command must be sent to the PIC to reset the in-service latch.

Programming the bits in the interrupt mask and how the EOI command is sent are discussed in chapter 4 where the software side of using interrupts is examined.

The interrupt sequence

Let's put all of this together and follow the sequence of events when a hardware interrupt occurs.

1. The I/O device in an adapter slot raises an IRQ line. (See note on the following page.)
2. The PIC responds by setting an internal latch.
3. The PIC checks priority, and if the request just received is the highest priority, the interrupt proceeds. Otherwise the request is held until higher-priority interrupts are done.
4. The PIC raises the INT pin on the CPU to request interrupt service.
5. If the CPU interrupt enable flag is cleared, the request is ignored. Usually, the interrupt flag will be set and the CPU will honor the request as soon as the instruction currently being executed is completed.
6. The CPU sends an –INTA (interrupt acknowledge) pulse to the PIC. The PIC freezes the current priority level, sets the in-service latch, and resets the IRR latch for this interrupt. (See note below.)
7. The CPU sends a second –INTA pulse to the PIC. The PIC then puts the interrupt vector number (a single byte) on the data bus.
8. The CPU uses the vector number to fetch the interrupt vector—that is, the address of the interrupt service rou-

tine (ISR)—from the table of interrupt vectors at the bottom of memory.

9. The CPU pushes the CS, IP, and flags onto the current stack and sets the interrupt flag.
10. The CPU jumps to the ISR and begins executing the code.
11. The next to last instruction in the ISR is an end-of-interrupt (EOI) command to reset the current interrupt in the PIC.
12. The last instruction in the ISR is an IRET (interrupt return). This causes the CPU to pop the flags, IP, and CS from the stack. The program code that was being executed when the interrupt occurred is resumed.

Note: There is a quirk in this process that can be important in some cases. The initial activation of the interrupt request occurs when an IRQ line is raised from low to high (it is positive edge-triggered), as in step 1. Unlike many edge-triggered devices, however, the PIC requires the IRQ line to remain high until the first –NTA pulse from the CPU (step 6). If the IRQ line goes low before the first –INTA, the requested interrupt will not be honored. Instead, a pseudo–IRQ7 is activated (but its ISR bit is *not* set, as it is when a genuine IRQ7 occurs). Chapter 5 has more to say about dealing with this quirk.

The fact that hardware interrupts are ranked in priority has practical consequences. A hardware interrupt will not necessarily interrupt the CPU immediately; the actual interrupt will not occur until all pending interrupts of higher priority have been serviced. The time between raising an IRQ line and the beginning of execution of the IRS is the *interrupt latency time*. The minimum time is set by the system design, including the clock rate and CPU type. As a very rough rule of thumb, the typical latency of an 80386 computer is on the order of 10 μs, a 286 AT is something like 15 μs, and an XT is 20 to 50 μs. Remember, this depends on the system design and is subject to considerable and unpredictable variation depending on what other interrupts might be active at the same time and whether any program code disables interrupts for a time.

Interrupts in the AT

The AT added seven more hardware interrupts to the eight interrupts in the OPC and XT by adding a second PIC. The PICs are connected in cascade—the output of one (the *slave*) is connected to one of the inputs of the other (the *master*). This provides a total of fifteen interrupt inputs. The basic idea is illustrated in Fig. 2-4, with some control interconnections between master and slave omitted for clarity.

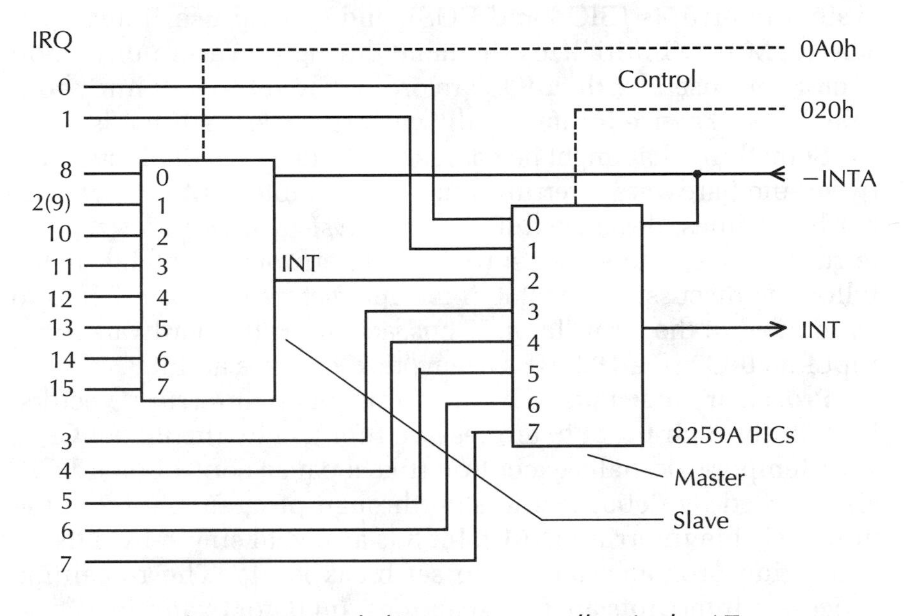

Fig. 2-4 *Cascaded interrupt controllers in the AT.*

In the AT the slave PIC is connected to the interrupt request 2 (IRQ2) input of the master. This means that the priority of the added interrupts is nestled among the older IRQ priorities. The newer IRQ8 through IRQ15 have a priority level less than IRQ0 and IRQ1 but greater than IRQ3 through IRQ7.

IRQ2 is a special case. The IRQ2 bus line was moved to input 9 on the slave PIC. For compatibility, what is really IRQ9 is redirected (revectored) by the BIOS to the same interrupt number (0A hex) used by IRQ2 in the OPC and XT.

As far as a program is concerned, interrupts originating from the master are handled in the same way as they are with the single PIC in the XT. In the case of the additional interrupts that go through the slave, however, there is a difference. The ISR must send two EOI commands: one to the slave and one to the master.

Standard interrupts

The interrupt vector table in the PC occupies the first 1024 bytes of memory (256 four-byte addresses). It's useful to divide them into four categories: processor interrupts, hardware interrupts, system interrupts (BIOS and DOS), and general user interrupts. The ROM BIOS initializes the table during power-up or reboot; vectors not used by the BIOS are ordinarily set to a dummy routine. DOS likewise initializes the vectors it uses as it loads.

Something that might be confusing at first is the distinction between the hardware interrupt lines and the interrupt vectors. The hardware lines, designated IRQ*x, are physical signal lines.* An interrupt *number* refers to an entry in the interrupt vector table. In the following discussion, the 256 interrupt vectors will be referred to by number in the form "Int *n*." Thus, activating the hardware interrupt line IRQ0 (the 18.2-Hz system "tick") evokes an Int 8.

Processor interrupts The first thirty-two interrupt vectors, Int 00h through Int 1Fh, are reserved for use by the 80*x*86 CPU. Int 1 temporarily halts execution (like a pause control on a VCR) and is used by debuggers to step through programs. Int 2 is the nonmaskable interrupt (NMI). Int 3 is a special single-byte INT a debugging program can use to set breakpoints. The remaining processor interrupts are for exceptions. Int 0, for example, is triggered by the CPU whenever division by zero is attempted.

Now the story gets messy. The 8088 uses only Ints 0 through 4 for its own purposes. Although the 8088 data sheet warns that other reserved interrupts might be used in future microprocessors, the designers of the OPC used many of these reserved but not yet implemented processor interrupt numbers for BIOS functions. Sure enough, when the 80286 and 80386 came along they did use more of them. Fortunately these chips are also sophisticated enough to allow a usable work-around.

Hardware interrupts The hardware interrupts are those administered by the interrupt controller. They are listed in order of priority. IRQ0, IRQ1, and (on the AT) IRQ13 are used by the system and do not appear on the bus. Table 2-3 lists the hardware interrupts.

Table 2-3 Hardware interrupt vectors.

IRQ level	Int number	Function
		Original PC and XT
IRQ0	08h	System timer
IRQ1	09h	Keyboard
IRQ2	0Ah	Unassigned
IRQ3	0Bh	COM2 serial port
IRQ4	0Ch	COM1 serial port
IRQ5	0Dh	Fixed disk in XT; unassigned in OPC
IRQ6	0Eh	Floppy disk controller
IRQ7	0Fh	Printer
		AT and 386
IRQ0	08h	System timer
IRQ1	09h	Keyboard
IRQ8	70h	CMOS real-time clock
IRQ2	0Ah	Unassigned (redirected from IRQ9 at 71h)
IRQ10	72h	Unassigned
IRQ11	73h	Unassigned
IRQ12	74h	Unassigned
IRQ13	75h	Numeric coprocessor
IRQ14	76h	Fixed disk controller
IRQ15	77h	Unassigned
IRQ3	0Bh	COM2 serial port
IRQ4	0Ch	COM1 serial port
IRQ5	0Dh	LPT2 printer
IRQ6	0Eh	Floppy disk controller
IRQ7	0Fh	LPT1 printer

System interrupts The major services provided by the ROM BIOS are accessed through software interrupts, primarily Ints 05h to 19h. A few have been redefined over time—Int 15h, for example, was the cassette interface in the original IBM PC—but for the most part, these assignments are valid for all PCs. Interrupts 20h through 3Fh are reserved for DOS. Programs use interrupts to communicate with DOS and to perform operating system functions (files, for example). The system interrupts are listed in Table 2-4.

Table 2-4 Standard software interrupt vectors.

Int number	Function
	Used by BIOS
10h	Video (all types)
11h	Configuration data
12h	Memory size
13h	Disk services
14h	Serial services
15h	Cassette in OPC; unused in XT; system services in AT
16h	Keyboard
17h	Printer
18h	ROM BASIC (IBM only)
19h	Disk boot
1Ah	System date and time
1Bh	Keyboard Ctrl-Break
1Ch	Timer tick (18.2 Hz)
1Dh	Video parameter table
1Eh	Disk parameter table
1Fh	CGA graphics characters
	Used by DOS
21h	General functions
23h	DOS Ctrl-Break handler
24h	Fatal error handler
2Fh	Multiplex
33h	Pointing device (mouse)
	Other
40h	Diskette revector
41h	Fixed disk parameters
42h–49h	Additional BIOS parameter tables
4Ah	User alarm (AT)
4Bh–5Fh	Reserved
67h	LIM EMS driver
70h	Real-time clock (AT)
75h	Coprocessor exception (AT)
80h–F0h	Reserved for BASIC in IBM PCs
60h–66h	User
F1h–FFh	User

User interrupts These are reserved for general use by application programs (that is, by you and me). Note that some adapter cards may install interrupt routines of their own in this area.

Data transfer methods

Few devices besides the keyboard produce data a few bytes at a time. More often the data comes in blocks or in streams. A substantial part of the art of I/O routines is dealing efficiently with such data, transferring it from a port into RAM rapidly and efficiently.

In dealing with block data it is important to distinguish between the average rate and the burst rate. The *burst rate* is the maximum rate at which the data flows during a block transfer. This might be considerably faster than the rate at which the application program can process data. All that is required is that the I/O routines manage to stuff it into a buffer in RAM as it comes in. The *average rate*, on the other hand, is simply the long-term average. The application program processing the data need only be fast enough to keep up with the average rate.

An example of the difference is the digital voltmeter project in chapter 6. It emits three five-digit readings per second, so the average rate is just fifteen bytes per second. But each reading is a block of five bytes at a (burst) rate of 600 bytes per second.

Interrupt-driven I/O

The most common way of handling data transfer is by means of interrupt-driven input routines. Each incoming byte triggers an interrupt; the ISR puts the incoming byte into a buffer set aside in RAM. (Outgoing data can likewise be handled by interrupts, of course, but polling is more common.)

Table 2-5 gives an idea of what can be expected from interrupt-driven I/O. The "conventional ISR" rate is for compact routines such as the serial port ISR given in chapter 4. The "parallel port" rate is for the eight-bit bidirectional printer port ISR given in chapter 5. Note that the assembly language versions of these routines were used to measure the timings. Please keep in mind that the values shown are approximate (though conservative) and are intended only as general guidelines. Different brands of computers of the same general type might give somewhat different results. The application being run might affect the rate also; for example, if it turns off interrupts for critical sections of code. Frequent interrupts of higher priority will likewise slow things down.

Table 2-5 Approximate interrupt-driven I/O throughput.

	16-MHz 386	8-MHz XT	4.77-MHz XT
Conventional ISR	34.5 kb/s	12.5 kb/s	8 kb/s
Parallel port, 8-bit	31 kb/s	10.5 kb/s	6.7 kb/s
Typical latency	15 μs	35 μs	50 μs

Looping

An interesting possibility is the transfer of data by looping. You fetch a string of *N* bytes by using a loop to read data from a port *N* times. An obvious flaw, and it is a fatal one, is that there is no way to guarantee that the next read will get the very next byte. The process is totally asynchronous. But there is an extension of this approach that will work and might be quite useful in some applications.

Suppose the port you are reading has some kind of internal buffer that holds incoming data. Suppose further that you can select an address in this buffer by writing a number—an index—to the port. The index is held in a latch, so that a read from the port fetches the data stored in the buffer at that index. A routine to move a block of data from the port to an array in memory could look like the pseudocode in Fig. 2-5.

Fig. 2-5 *Pseudocode for polling an array in a device.*

```
for 1 to Count do
   begin
      write index to port
      read result from port and put into Array[N]
      increment index
      increment N
   end
```

(In an actual routine it would be better to do the index increment between the write and read, more nearly balancing the cycle times.) There would have to be some means of signaling when the buffer is full and ready to be read. This could be done by an interrupt, or by setting a status byte (at a different port address) that the program could poll.

It is not difficult to design a custom I/O board using this principle that can plug into an adapter slot in the computer. Inexpensive static RAM could serve as the buffer. Decoding address lines A0 to A9 plus AEN would provide a "card selected" signal. A low –IOW, indicating a write to the card, would take the index from the data bus and store it in a latch. A low –IOR, indicating a read, would enable a driver that would place the data stored in the buffer on the data bus. A counter, clocked by the incoming data strobe, would keep track of the amount of data in the buffer. Working out the design is left as a project for interested readers, partly because it depends on the needs of a particular application, and partly because our main concern in this book is I/O without plug-in accessory adapter cards.

A variation on this technique is of considerable interest, however. As will be seen in chapter 5, the standard parallel (printer) port can be manipulated in software so as to give bidrectional, interrupt-driven I/O. The looping technique can be put to work as a simple and efficient way to read a remote buffer. Let me give an actual example. An interface was constructed for use with a multichannel digital recorder whose output data stream consisted of eight sixteen-bit words, one for each channel, at a rate of 11,025 words per second. The application required data at a rate of only 300 values per second. The interface was connected to a PC through its parallel port. A timer in the interface put a block of eight words (as sixteen-byte pairs) in a small buffer every 3.3 ms and then wrote a status byte to the computer, triggering an interrupt. The ISR then looped through sixteen addresses in the buffer, moving the bytes from the interface into memory in the computer. Note that the tactic used was to combine an interrupt to initiate a transfer with looping for the actual transfer. This is a powerful and useful technique.

The example nicely illustrates some of the more general points this book tries to make. This simple interface cost about $75, as compared to the $600 plug-in adapter card offered by the recorder manufacturer. Moreover, the application sometimes required setting up in the field with a laptop computer, which wouldn't accept the commercial card in any case.

Table 2-6 is offered as a rough guide of performance to be expected from the looping technique. The "parallel port" rate is for the eight-bit bidirectional printer port I/O routine described in chapter 5. The values shown are the burst rate; that is, the rate achieved during the transfer loop itself. A block size of 256 bytes was used for these tests. Please bear in mind, again, that these rates are only approximate.

Table 2-6 Approximate looping I/O throughput (burst rate for 256-byte block).

	16-MHz 386	8-MHz XT	4.77-MHz / XT
Custom I/O board	155 kb/s	99 kb/s	66 kb/s
Parallel port, 8-bit	91 kb/s	40 kb/s	25 kb/s

These two looping techniques underscore the point made earlier that I/O transfer speed depends on more than raw CPU speed. Figure 2-6 plots the speed of the simple custom I/O board and the parallel port routine as a function of CPU execution speed relative to a 4.77-MHz XT. Note how diminishing returns clearly set in. The dotted lines show what would result if CPU speed and the I/O transfer rate were linearly proportional. (The curves shown are calculated, but measurements made on a variety of machines were in good agreement.)

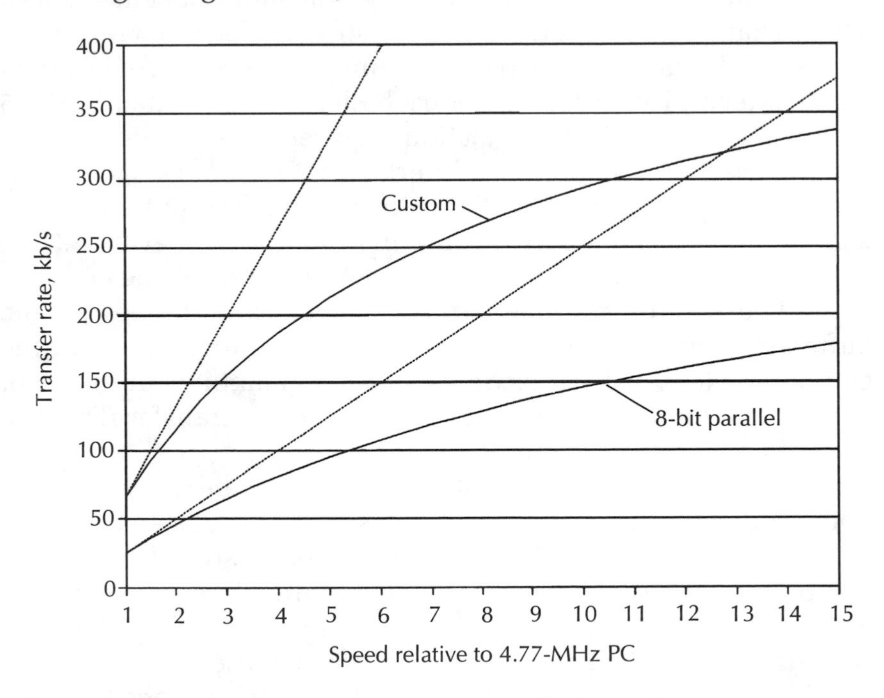

Fig. 2-6 *Relative looping throughput.*

Direct memory access

Although careful use of the I/O channel can achieve fairly good data-transfer rates, it seems we ought to be able to do even better. For one thing, it seems a little roundabout to move data between I/O ports and memory by going through the CPU registers. This is necessary for moving data between two locations in memory because the memory bus cannot be in the read state and the write state at the same time. But the states of the I/O bus and the memory bus are independent of one another. Moving a byte directly between a port and a location in RAM in one step should be possible, and certainly more efficient. Similarly, sequential transfers would be faster if addresses could be incremented in a hardware counter rather than by CPU arithmetic. Many computers, including the PC, do provide just such a function—direct memory access (DMA).

The 80*x*86 is designed to share the system busses with other processors. During DMA operations, a DMA controller takes over the bus and the 80*x*86 remains idle. The DMA controller is in effect a very specialized processor. The Intel 8237A used in PCs has four independent channels. In the PC and XT, channel 0 is used for memory refreshing, channel 2 for floppy disk service, and a channel 3 for servicing a hard disk if one is installed. Channel 1 is free for general use. More channels are available in the AT, which uses two controllers in cascade and has separate hardware for memory refreshing.

Each channel has a count register that holds the number of bytes to be transferred (up to 64K) and a sixteen-bit destination address counter. (A small additional static RAM is added to extend this to twenty bits in the PC and XT and twenty-four bits in the AT.) Each channel also has several control registers that select the mode of operation and whether the transfer is into memory (write) or out of memory (read). There is a request input line (DRQx) for each channel that a device can use to request a DMA transfer and an acknowledge output line (DACKn) for "handshaking" with the requesting device.

A somewhat simplified picture of a DMA transfer in the PC is as follows. When a request is granted by the DMA controller, the controller asks the CPU for use of the bus. When the CPU gives the bus over to the DMA controller, the DMA cycle begins. The DMA controller sets the address bus to the value in the destination address counter and sets the destination bus state to read or write. It then sends an acknowledge signal (–DACK) to the device requesting the DMA. The device then puts data on the bus (for write) or receives it from the bus (for read). When the cycle is complete, the controller tells the CPU to proceed and also decrements the count register and increments the address register for that channel. If the count register in the controller has not reached zero, the controller initiates another request. Such single-byte DMA cycles continue until the count register in the controller reaches zero (actually, one count beyond zero). The DMA controller then activates the –TC (terminal count) line, indicating to the requesting device that the preprogrammed number of bytes have been transferred.

Although the 8237A can be set to transfer up to 64K in one operation, that mode is not used in the PC because it could interfere with RAM refreshing. Only the single-byte mode outlined previously can be used, allowing the CPU and the DMA controller to take turns on the system bus. Even so, rather fast data rates can be realized (as much as 450 kb/s in a 4.77-MHz PC).

Any I/O device that wants to use DMA must be specifically designed for that function, use the single-byte transfer mode, and have the appropriate hardware to interface with the DMA controller lines. Because the focus of this book is on general-purpose I/O, and especially on the use of the PC's built-in I/O ports, further details of DMA operation are beyond its scope.

Operating system I/O

Although this book is about direct external I/O, most every program also uses the DOS I/O system, for files if nothing else. Beginning with version 2, DOS classifies all the I/O devices with which it deals as either character or block devices. *Block devices* include disk drives and the like which handle blocks of data. *Character devices*, such as the keyboard, the display, serial ports, and printers, are assumed to operate a byte at a time. Which category a given device belongs to is specified by the driver for that device.

DOS very conveniently treats block and character I/O devices the same way—namely, as a file. But there is a crucial internal difference. The data stream to and from block devices is handled "as is" without any manipulation or filtering. However, DOS ordinarily buffers character devices and inspects each byte to see if it is a special character such as ^C (break) or ^Z (end of file) and acts accordingly. This can cause unexpected and vexing problems. It is common nowadays for output to the printer to include graphics data as well as text (ASCII) characters. If a graphics data byte happens to have the value 26, DOS will interpret it as a ^Z end-of-file marker and unceremoniously terminate the printing. One way to avoid this is to bypass DOS and use the ROM BIOS directly. In this case a program must have two sets of output routines in order to allow printer output to be directed either to the printer or to a disk file.

A more elegant method is to tell DOS not to filter the stream from a character device—in DOS jargon, to handle it in "raw" rather than "cooked" mode. This can be done with the DOS IOCTL (I/O Control) function.

Few programming languages appear to support IOCTL calls directly, but it is easy to write your own if the language provides access to CPU registers and file handles. Two useful routines are shown in Fig. 2-7. CharDevice returns true if a specified file is a character device, or false otherwise. SetCharMode allows a character device file to be set to either "raw" or "cooked" mode. Calling SetCharMode with raw set to true forces raw mode; with raw set to false, cooked mode is set.

There are two points to note in using these routines. First, the file in question must already be open. Second, the file must be designated by its DOS "handle," a sixteen-bit word assigned by

Fig. 2-7 *DOS IOCTL routines.*

```
USES Dos;

{ Check if file pointed to by Handle is a character device. }

function CharDevice( Handle : word ) : boolean;         { TRUE if char device }
var
   Regs : Registers;
begin
   Regs.AX := $4400;                                   { get IOCTL for handle H }
   Regs.BX := Handle;
   MsDos(Regs);
   CharDevice := (Regs.DL and $80 = $80);
end;


{ Set Raw mode (Raw TRUE) or cooked mode (Raw FALSE). }

procedure SetCharMode( Handle : word; Raw : boolean );
var
   Regs : Registers;
begin
   Regs.AX := $4400;                                   { get IOCTL for handle H }
   Regs.BX := Handle;
   MsDos(Regs);
   if Regs.DL and $80 = $80 then                   { if handle is a char device.. }
      begin
         Regs.DH := 0;
         if Raw then Regs.DL := Regs.DL or $20           { set bit 5 for raw }
            else Regs.DL := Regs.DL xor $20;        { clear bit 5 for cooked }
         Regs.AX := $4401;
         Regs.BX := Handle;
         MsDos(Regs);
      end;
end;
```

DOS when the file is opened. In Turbo Pascal (4.0 and later) this is kept in the internal control structure maintained for each file. If "MyFile" is a file of type Text, the handle is available as TextRec(MyFile).Handle; for all other file types it is FileRec(MyFile).Handle. (In Turbo Pascal 3.0 you must access the file record indirectly by declaring a variable on top of it; for example, Handle : integer absolute MyFile;. You can then pass Handle to SetCharMode. Turbo 2.0 and earlier do not use DOS handle calls.)

DOS provides very fast and efficient unfiltered reads and writes of block devices such as disk files. Such transfers are done by calling subfunction 3Fh (read) or 40h (write) through the DOS interrupt (Int 21h). Although most programming languages use this method internally, their ordinary read and write functions typically add a layer of further manipulation, making them less efficient. A number of language compilers for the PC do provide the option of direct high-level calls to this function; for example, BlockRead and BlockWrite in Turbo Pascal, and _read and _write in Turbo C.

Summary

This chapter has examined some of the inner workings of the PC, particularly the I/O functions. Much of what has been covered is background for the topics of the chapters to follow. Two areas are especially important: the I/O ports with their addresses and interrupt numbers and the interrupt system itself. I will be drawing constantly on these two for the remainder of this book.

❖3 Principles of data transmission

This chapter considers data transmission. This major area of engineering might seem out of place in a book on data acquisition, but not so. After all, getting information into and out of a computer necessarily involves data transmission. Another reason to look at this topic is to correct some common misunderstandings of data transmission methods, especially in the case of the widely used RS-232 serial format. The reputation that serial transmission has for being quirky is largely undeserved.

Let's start at the beginning. All transmission systems can be classified as parallel or serial. These are very general terms. They are also somewhat relative—all the bits of a byte are sent simultaneously in a parallel system, but the bytes themselves are sent serially. Within this broad division, many variations are possible. For example, the 8088 microprocessor handles sixteen-bit I/O as two sequential eight-bit bytes. On the other hand, certain error-detection schemes regard a data block or an entire file as one gigantic *n*-bit word. "Serial" and "parallel" are words that get much of their meaning from the particular context in which they are used.

To describe a transmission system fully, three more characteristics must be specified: (1) the nature of the physical (electrical) signals; (2) the format of the data; and (3) the means of transmission control. Each of these things will be considered when looking at the transmission methods most often used in data acquisition.

A general conceptual scheme or model for data communication has been worked out and standardized under the International Standards Organization (ISO). It divides the process of data transfer into a set of *levels*. Level 1 is the "physical" interface: the nuts and bolts (or volts and Hertz) of getting signals from one place to another. Level 2 is the "link control," which concerns the format of the data that is transmitted and any protocols used for controlling flow. Level 3 is the "network control" level, which supervises the transfer. Although the ISO model is meant specifically for telecommunication networks, its principles are very helpful in clarifying our thinking about data systems. A case in point is the EIA RS-232 standard for serial data interchange. There seems to be a fairly widespread misunderstanding that it specifies a particular serial transmission format—so-called asynchronous serial. But not so. It addresses level 1, the physical level. IBM offered two "synchronous" serial adapters for the PC—SDLC and IBM Bisync—and both of them used the RS-232 interface.

Because several modern transmission systems still show vestiges of their origins in telegraph and teleprinter technology, let's begin with a glimpse at history.

Background

The idea of using electricity to communicate seems to go back at least to the eighteenth century, but to become practical it had to wait for the steady and reliable supplies of current developed in the nineteenth century. The early systems were based on the transmission of written text. One of the first systems appears to have been a one-wire-per-letter scheme devised in 1810 by von Sömmering. This was very inefficient and various methods of coding letters to use fewer wires were developed. In 1839 Charles Wheatstone (of resistance bridge fame) and W. Cooke devised a double-two-of-five magnetic system that, though it provided only twenty letters, was actually used for a time.

The first genuinely successful system, the telegraph, was a strictly serial system that required only one transmission circuit. Samuel F.B. Morse did not invent telegraphic signaling, but he does deserve credit for pioneering telegraphy—that is, a telegraphic system with software (his code) as well as hardware. His original receiving apparatus was a graphic device using a sheet of paper attached to a rotating drum. Current flow activated a solenoid that caused a stylus to mark the paper. This led to a terminology still used in telecommunications: "mark" for the high or on level of a circuit, and "space" for the low or off level.

The Morse telegraph has some subtleties whose significance was not fully appreciated until later times. Although the telegraph is an on-off system, it is not a binary but a three-state (ternary) system: dot, dash, and silent. In modern terms it is a combination of two modulation methods: on-off (pulse amplitude) and duration (pulse width). Combining amplitude and time (phase) modulation is how modern high-speed modems pump so much data through ordinary voice-grade telephone lines. It also used a form of data compression, not unlike Huffman coding. The number of dots and dashes in each character was made roughly inversely proportional to its frequency in ordinary English; for example, the letter E is just a single dot.

These rather sophisticated techniques required sophisticated encoders and decoders—human telegraphers. A number of attempts were made to fashion a wholly mechanized system. The first successful system was one devised by J. M. E. Baudot in France in the 1870s. It simplified things by using a uniform bit length (no dot versus dash), a uniform character code length, and pure binary (on-off) coding. Baudot worked out a five-level code, a modified version of which is still standard for electromechanical teleprinters. Five bits provides only thirty-two characters, so two "shift" characters are used, FIGS to switch the interpretation to the "shifted" set, and LTRS to return to unshifted. (This is the origin of the shift in (SI) and shift out (SO) characters in the ASCII code.)

Thus was born the teleprinter or teletypewriter (TTY) and with it a vast and productive technology. The TTY transmission format has proved itself to be remarkably robust—with a number of important modifications, it is still one of the chief methods of serial data transfer. It is the format used by the PC serial port.

Parallel transmission

Parallel systems are simpler than serial systems. They are also inherently faster for a given bit rate, because entire groups of bits are sent together rather than just one bit at a time. The price that is paid for these advantages is that more transmission channels (wires, tape tracks, and so on) are required.

The simplest parallel system is a group of wires, one per bit, which together represent a data byte. Such a system is a blind parallel system in the sense that the data is just there in the lines without any way for a device to tell when the data changes or whether it is currently valid. Blind parallel systems are not without value. There are many applications in which only the current value of the data is important; a digital voltmeter or frequency counter, for example.

In many other cases this will not do. There needs to be some means of signaling when the data is valid, when it has changed, and so forth. Parallel systems commonly contain one or more signaling lines in addition to the data lines. The simplest form is the status line in which the line changes state to indicate that the data lines contain valid information.

An important use of a signal line is to avoid transmission errors. In practice there is usually some noise and distortion during signal transitions, and in many cases the individual lines might not change state simultaneously. This is illustrated in Fig. 3-1. To avoid these sources of errors, parallel systems are often run with *data strobes*, an additional pulse that is narrower than the data pulses. The strobe begins after the data lines have had time to settle and ends before the transition to new data. The receiving device is designed to latch the data during the strobe.

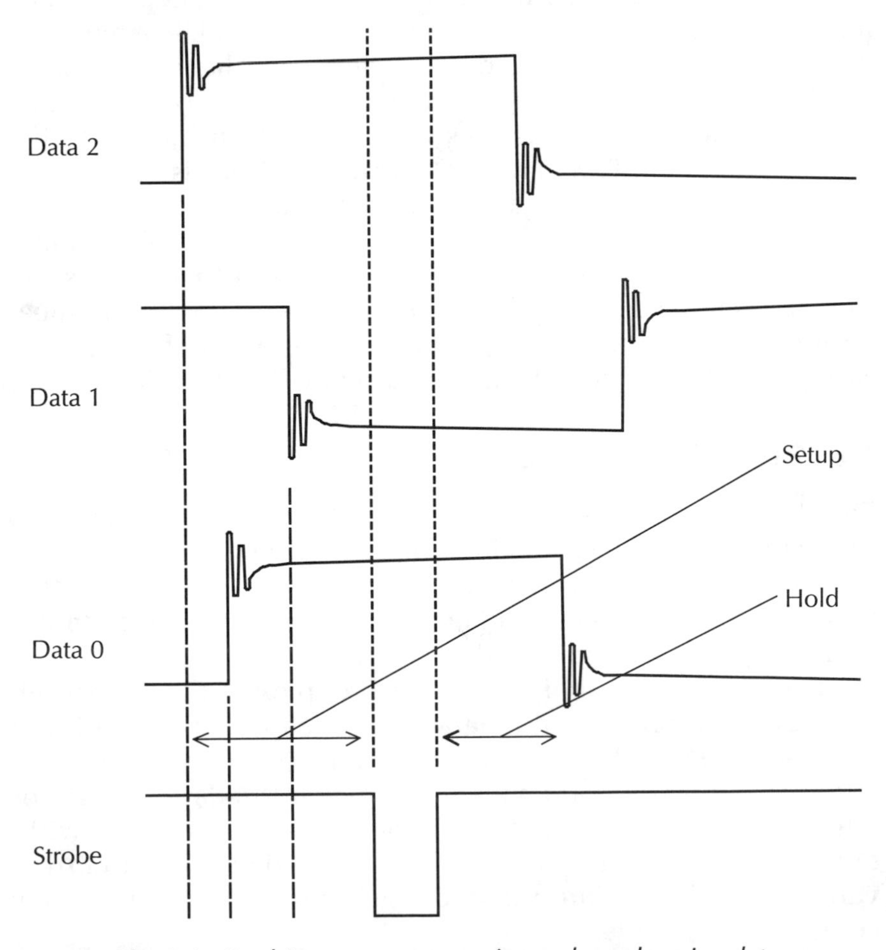

Fig. 3-1 *Strobing to suppress noise and synchronize data.*

The Centronics interface

A familiar and widely used parallel transmission system is the printer interface originally developed by Centronics for their line of printers. As shown in Fig. 3-2, the path from the computer to the printer consists of eight data lines and a strobe. There is also a path from the printer to the computer consisting of an acknowledge pulse and several status lines. The most important status line is BUSY, which goes high when the printer is not ready for more data. There are other status lines for paper out, errors, and selected (on-line or off-line). Electrically, all signals are standard TTL level and the lines are essentially unterminated. (Line termination is discussed in chapter 7.) This limits the Centronics interface to rather short cable distances, generally 12 feet or less.

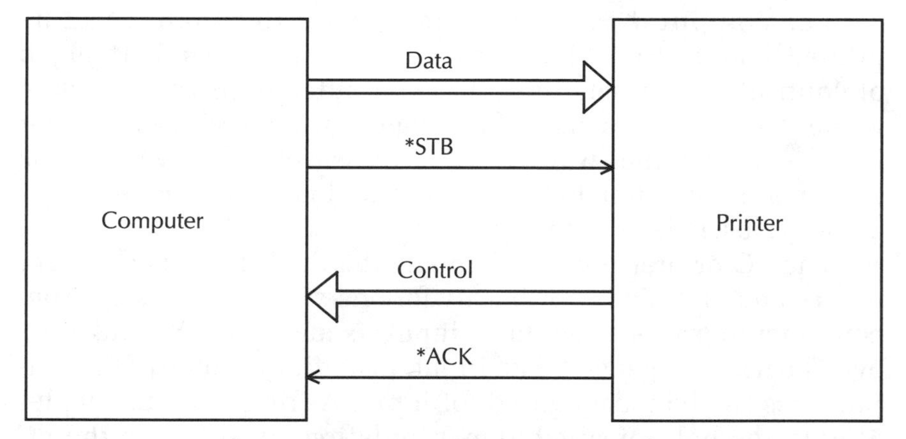

Fig. 3-2 *The Centronics printer interface.*

Figure 3-3 shows the signals in the data path from computer to printer as well as the *ACK pulse and the BUSY status signal. A character to be printed is put on the data lines, and after a setup time of 1 μs or so the *STB pulse (typically 1 to 2 μs wide) strobes the data into the printer. The printer raises BUSY while the data is being entered into its buffer, then drops BUSY and issues an *ACK pulse to say "data accepted; go ahead." (For reasons that are discussed in chapter 5, the *ACK signal from the printer is hardly ever used by PCs.) This cycle repeats as each character is sent to the printer. Notice that the rate of data flow is not fixed but variable, and that it is controlled jointly by the computer and the printer. This method of mutual control is widely known as *handshaking.*

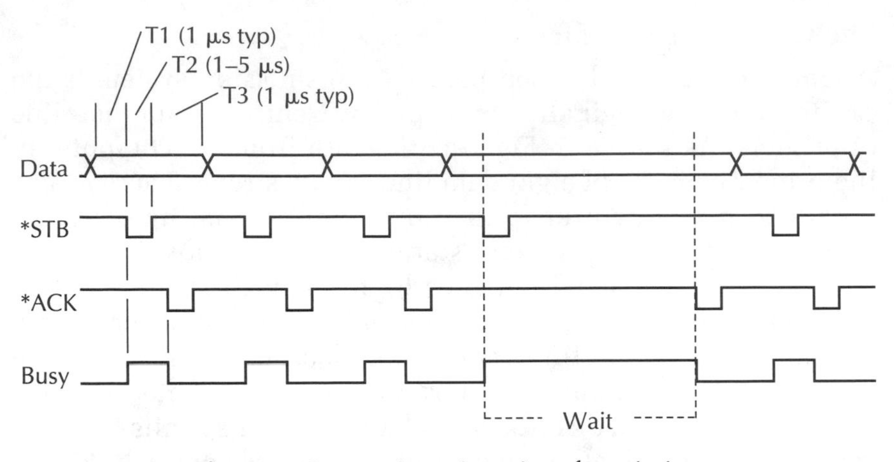

Fig. 3-3 *Centronics printer interface timing.*

The computer can send data much faster than it can be printed, and the BUSY line from the printer has the important job of controlling the data flow. The computer stops sending data while BUSY remains high—for instance, when the printer buffer is full—and resumes when BUSY returns to low. Flow control is an important function in a great many systems and will be looked at more closely later in this chapter.

The PC printer port implements the Centronics parallel interface and is designed primarily for one-way transmission from computer to printer. The status inputs (such as BUSY) can, however, be used for general data input. In addition, some of the output lines can be reconfigured for input. As discussed in chapter 5, with the help of suitable manipulations in software the PC printer port can also be used quite effectively for general-purpose parallel input.

The GPIB system

Next to the printer interface, the most widely used parallel PC interface is undoubtedly the general-purpose instrumentation bus (GPIB). It is intended specifically for data interchange among groups of up to fifteen instruments in data acquisition and instrumentation systems. It was designed in the mid-1960s by Hewlett-Packard, but it has since been standardized by the Institute of Electrical and Electronic Engineers (IEEE) (standard 488, *Standard Digital Interface for Programmable Instrumentation;* the current revision is 488.2) and it is also known as the "IEEE 488" interface.

The GPIB is a good deal more than an interface; it is a complete transmission and control system. A detailed description would run to many pages, and only a brief summary is given here. For more information, see one of the introductions to the GPIB such as Caristi (1989).

The GPIB is fully bidirectional. Devices can put data on the bus (be a "talker" in GPIB jargon) or receive data from it (be a "listener"). Any device can exchange data with any other. Each system has a controller that serves as system supervisor and traffic cop. PCs are often used for this function. Each device has its own system address, set either by software or by switches in the unit itself.

A GPIB system can be set up in several ways. An interesting feature is that the standard connector is specified as a totem pole plug-and-socket combination (as on Christmas tree lights). As shown in Fig. 3-4, this allows various configurations to be put together quite simply.

The bus itself (Fig. 3-5) consists of eight data lines, five control lines, and three flow control lines. All are open-collector or three-state TTL lines, with a standard termination network on each connector. Negative logic is used, with 0 as high and 1 as low. In principle, the bus is capable of reasonably high speeds; some vendors claim as high as 1 Mb/s. This is misleading, though, because the bus speed varies depending on which devices are using it at any moment. In fact, it always runs at the speed of the slowest of the currently active devices.

The GPIB uses a unique three-line handshake that combines aspects of the STB/ACK signals and the BUSY line methods. The three handshake lines are NRFD, DAV, and NDAC (see Fig. 3-5).

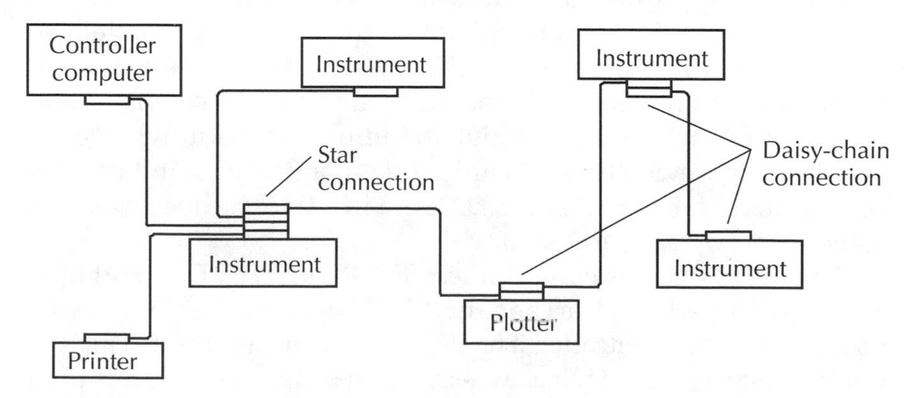

Fig. 3-4 *A typical GPIB instrumentation setup.*

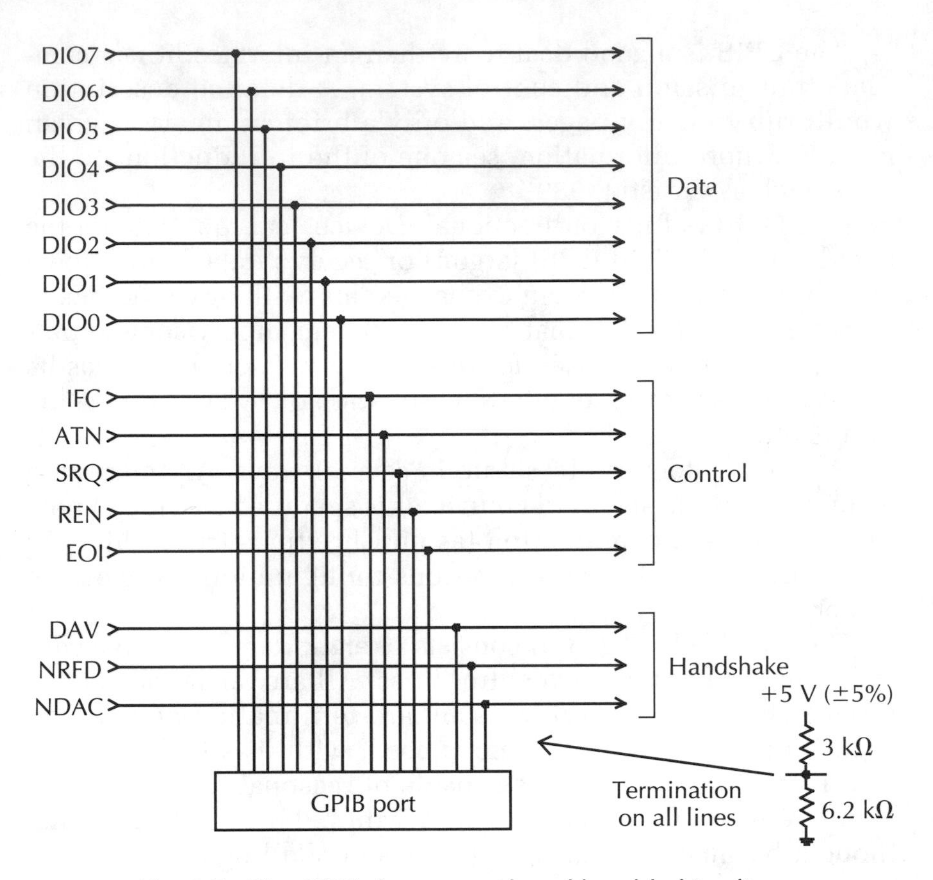

Fig. 3-5 *The GPIB data, control, and handshaking lines.*

Whenever any device is not ready to receive for any reason, it pulls down the NRFD (not ready for data) line. Because all devices are in parallel, the whole system is put into a BUSY state. The DAV (data valid) line is the system data strobe. When the bus is strobed, each device immediately pulls the NDAC (no data accepted) line low and releases it only after it has read the data lines and gotten the data byte. Only after all devices have read the data lines will NDAC go high to indicate an acknowledge. No transmitting device (talker) can put data on the bus and activate DAV unless both NRFD and NDAC are inactive, indicating all devices are now ready to listen.

There are five bus control lines: IFC, REN, ATN, EOI, and SRQ. The "IFC"(interface clear) and the "REN" (remote enable) lines are used only by the controller. The IFC line is a master reset. The REN line essentially tells devices whether to use the bus or revert to remote (stand-alone) operation. The "ATN"(attention) line is used to distinguish system commands (ATN on) from data (ATN off). The "EOI" (end or identify) line has two uses. When data is being transmitted (ATN off), the EOI line indicates the end of the data. When the system is in command mode (ATN on), the EOI line initiates a parallel poll of the devices. The "SRQ" (service request) line is like an interrupt: a device activates it whenever it needs attention for any reason (for example, out of range, error, and so on).

Devices are controlled by commands sent over the data lines with the bus set for command mode (ATN on). There are basically three categories of commands. There is a small set of universal commands that all GPIB interfaces must recognize. There is a larger set of general-purpose commands, not all of which are used by a given device. Finally there are the commands specific to each device that are set by its manufacturer.

The GPIB is exceedingly well suited for setting up flexible measurement, process control, and testing systems involving several instruments. Most modern instruments are available with GPIB interface, and a number of control and display programs are available that make programming a GPIB setup and analyzing results fairly simple. Its disadvantages are the need to use special cabling and to install GPIB interface cards in PCs that will be used with the system. This adds to the cost of a system; on top of that, most instrument prices are higher for GPIB interfaces than for RS-232 serial.

Flow control

Data flow control is an important part of any transmission system. Very often the sender and the receiver differ in the speed with which they can process data, and some way has to be provided for the receiver to initiate a pause while it catches up or data will be lost. Errors and other occurrences (such as a printer running out of paper) should also have a way of suspending data traffic.

Flow controls in the Centronics interface and the GPIB use separate lines for control, and are examples of hardware flow control. The simple kind of flow control in the Centronics interface using the BUSY line is the most common form of hardware flow control and is often referred to as RDY/BSY (ready/busy) flow control. RDY/BSY control can be used in serial interfaces also, and is the method often used by printers and plotters.

A different approach to data flow control uses special codes embedded in the data stream itself. Such methods are, not surprisingly, known as software flow control. When only a single transmission channel is available, as in the case of telephone links using modems, software flow control is the only method possible. Although many people associate software flow control with serial systems, it is perfectly usable in parallel systems also. Most parallel systems do in fact use hardware flow control.

The most common software flow control method is the XON/XOFF system. When the receiving device can't accept any more data it sends the XOFF character to the transmitter. When it is ready to continue, it sends the XON character. By convention, XON is ^Q (17 decimal, 11 hex), also known by its ASCII mnemonic **DC1**. XOFF is ^S (19 decimal, 13 hex), ASCII **DC3**.

Another common method is ETX/ACK. The originating system sends data in blocks and appends the ASCII end-of-text character **ETX** (3 decimal, 03 hex; ^C) to each block. When the receiver is ready for the next block of data, it emits the ASCII acknowledge code **ACK** (6 decimal, 06 hex; ^F). A disadvantage of this method is that the receiver has no way of stopping transmission before the end of the block in the event of an error or other unexpected condition. The sending system must also know the maximum block length that the receiver can accept.

One limitation of software flow control is that there has to be a way to distinguish control bytes from data bytes. The RDY/BSY and ETX/ACK methods come from ASCII character transmission practice in which byte values 0 to 31 are reserved for control functions. Some sort of work-around has to be used when such flow control is used with full eight-bit data. One method is to separate control and data bytes by position. For example, the ASCII codes **STX** (start-of-text, ^B) and **ETX** can be used to mark a block within which all bytes are data bytes. Another method is to manipulate 8-bit data so that the values 0 to 31 never occur. The Kermit file-transfer protocol developed at Columbia University uses a combination of bit shuffling and prefix codes to restrict the data to values between 32 and 126, and further delimits data to blocks of 96 or fewer bytes. All of this fancy footwork makes Kermit slower than many other protocols, but it is also one of the most "bulletproof" protocols ever devised.

An important consideration in system flow control design and operation is the *latency* of the control—the time it takes for a control signal to be recognized and acted on. The ideal latency would be zero. Hardware flow control is inherently fast and for most ordinary I/O applications can be regarded as having close to zero latency. Software flow control using embedded flow control codes over a slow transmission link might have considerable latency. Whatever the minimum possible latency of the transmission medium, in practice there will be some additional latency because it takes the receiving system a certain amount of time to recognize the flow control signal and some additional time to respond to it. For these reasons it is common to build in some leeway in systems using flow control. A receiving device might issue an XOFF when its buffer is, say, three-quarters full rather than waiting until it is completely full.

Serial transmission

Serial transmission systems take each byte and transmit or receive its constituent bits sequentially over a single circuit. This can be done in many ways, and there are lots of different (and incompatible) serial transmission systems in general use.

At the heart of all serial transmission systems is a means of converting byte data to sequential bit streams, and vice versa. Such conversions are easily done with a shift register. Figure 3-6 illustrates the principle. Shift registers are usually implemented as a cascade of D flip-flops clocked together. The data input of each flip-flop is the data output of the previous one, so the data bits are transferred left to right on each clock pulse. It's like a bucket brigade. For parallel-to-serial conversion (Fig. 3-6a), the register is loaded by using each bit to control the SET input of its corresponding flip-flop. The rightmost output becomes the serial stream. Note that the input byte has its least significant bit to the right, so it is shifted out lsb first and msb last. Many serial systems, including the PC serial port, follow this convention, but others use the reverse order with the msb first.

Conversion in the other direction, serial to parallel, is shown in Fig. 3-6b. The serial stream goes to the data input of the leftmost flip-flop and the stream is clocked through. After the appropriate number of clocks—eight in this example—the data outputs of each flip-flop will contain the values of each bit. The two shift register configurations are obviously quite similar. Many shift register chips (such as the LS/HC165 and the CD4021) contain internal gating to allow them to be used either way.

The output from an eight-bit byte-to-bit conversion is shown in Fig. 3-7. This is the "natural" and most basic serial format, commonly called nonreturn to zero (NRZ). The slightly odd name comes from the fact that a long string of 1s would keep the serial signal at high level continuously. (The name is sometimes

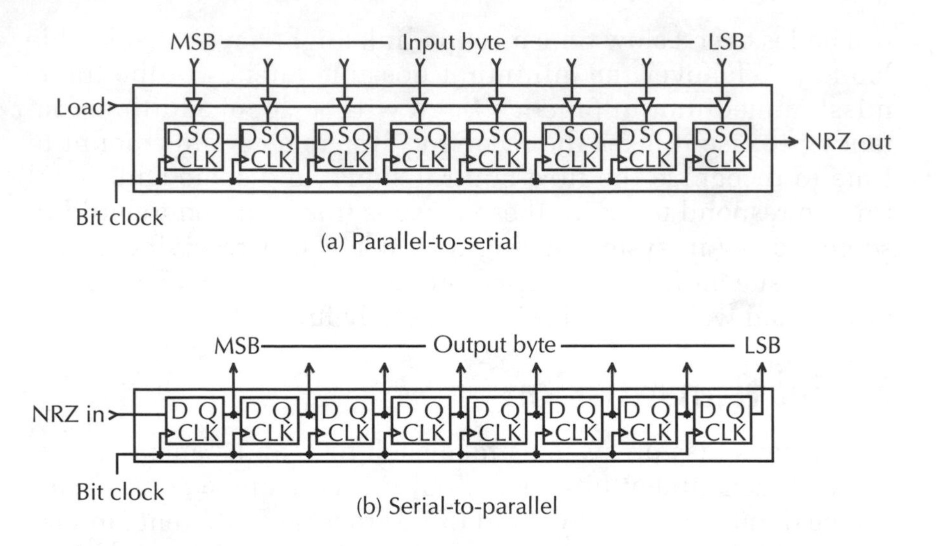

Fig. 3-6 *Parallel-serial (a) and serial-parallel (b) conversions.*

used carelessly to refer to differential NRZ in which 1s produce a signal transition and 0s do not.)

On the receiving end, the NRZ signal is fed into a shift register connected for bit-to-byte conversion to recover the data byte. In practical systems the serial waveform is apt to be somewhat rounded and perhaps noisy by the time it gets to the receiver, and a latch is usually placed between the input and the shift register so that the input is sampled at the midpoint of each bit (shown by the arrows in Fig. 3-7).

By the way, the video output from the monochrome, CGA, and EGA video cards is a TTL-level NRZ serial signal. The transmission rate is 14 to 16 megabits per second.

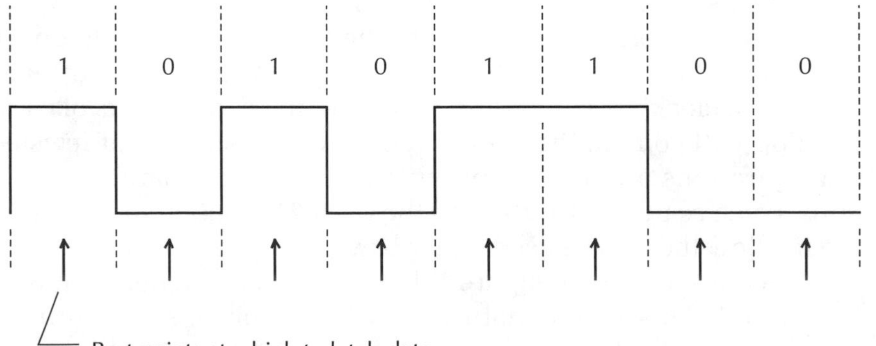

Fig. 3-7 *NRZ serial coding.*

Signal bandwidth

The NRZ serial format is simple and efficient, but it has several characteristics that preclude its use in some cases. Long strings of all 0s or all 1s will keep the signal at space or mark level for quite a while. The signal becomes, essentially, dc. This is of no consequence in some systems, but it is fatal in others. Systems such as telephone networks that send signals through transformers cannot pass dc and very low frequencies. The read heads on disk and tape drives respond to the rate of change of the magnetic signal and likewise have a low-frequency limit.

One solution is to use the NRZ signal to modulate an ac carrier. For example, a mark level could turn the carrier on, and a space would turn it off—which is exactly what is done in radiotelegraphy. Better and more sophisticated methods use frequency, phase, or amplitude modulation of the carrier, or some combination. This approach is used in the modem (*mod*ulator-*dem*odulator).

Another approach is to change the electrical format of the line coding in such a way as to avoid low-frequency problems. A method widely used in large telephone networks is alternate mark inversion (AMI) coding, shown in Fig. 3-8. Each mark is transmitted with a polarity opposite to the last mark transmitted; spaces, being zero already, are transmitted as is. Because the average value over several bits is zero, it has no dc component. An interesting and important property of this code is that, unlike most codes, it is insensitive to phase inversion. A similar idea is used in the bipolar RZ (return-to-zero) code also shown in Fig. 3-8. Each of these codes is easily formed from an NRZ data stream.

Synchronization

Another and potentially more serious problem with NRZ coding is that it requires the transmitter and receiver clocks to be synchronized. If they are not, they will drift apart and data will be recovered erroneously. Except in rare cases where it is practical to provide a second circuit to carry clock information, NRZ serial systems have to rely on independently generated transmitter and receiver clocks.

If the line coding is such that there is a signal transition (low-high or high-low) at least once in every single bit cell, the data stream itself would implicitly contain clock information. It would be self-clocking. Several such codes have been devised. Some are based on frequency modulation (FM). In the simple FM code in Fig. 3-9, widely used in the early days of microcomputers for recording

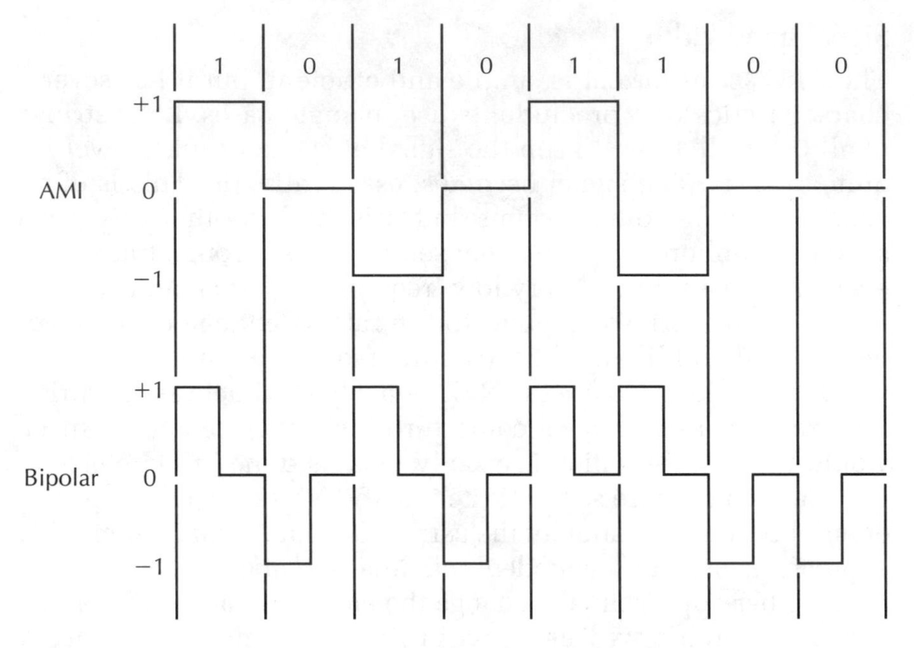

Fig. 3-8 AMI and bipolar serial coding.

data on audiocassettes, the coding switches between two frequencies—in the figure, Fc for zeroes and 2Fc for ones. This scheme is known generally as frequency-shift keying (FSK) and it is widely used in ham radio RTTY links and the standard Bell 103 300-baud modem. A variant is the pulse-position FM code, which in effect interleaves clock and data pulses.

An excellent and widely used code is the biphase code shown in Fig. 3-10. It uses phase rather than frequency modulation. Sup-

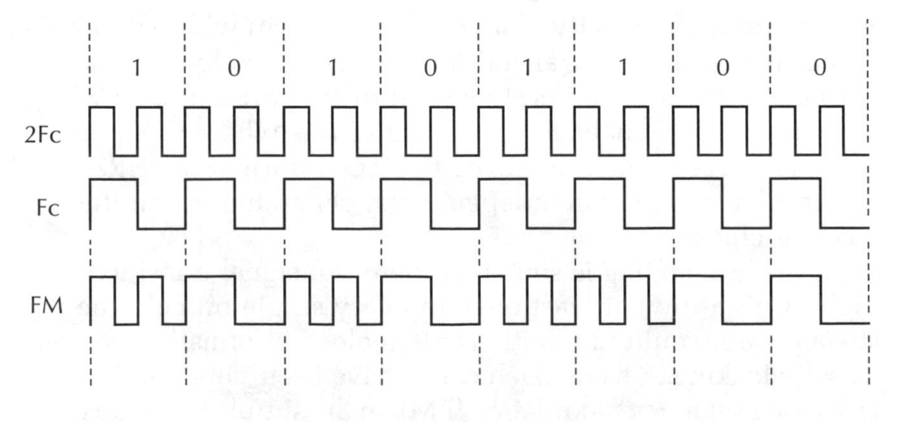

Fig. 3-9 FM serial coding.

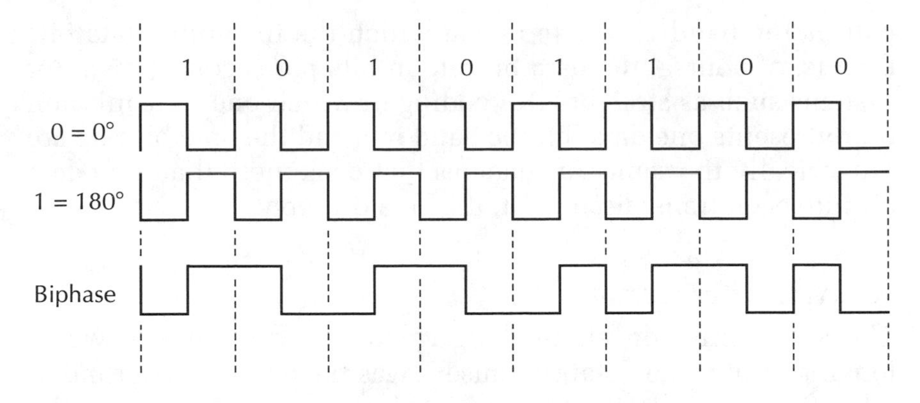

Fig. 3-10 *Biphase serial coding.*

pose the bit clock at 0° phase angle is regarded as a mark; space is then the clock at 180° phase. Note that a positive-going transition in the middle of a bit cell will always be a mark, and a negative-going transition will always be a space. Synchronization is readily accomplished by sending several zero bits for frequency locking followed by a couple of purposely miscoded bits (for example, high for 1½-bit periods) to establish a phase reference.

The principle can be extended to phase shifts other than 0° and 180°. A phase shift in multiples of 90° could be used. This would provide four distinct states—0°, 90°, 180°, and 270°. Such an approach is used in the familiar 212A (1200 bps) modem, in which the carrier is phase modulated in 90° increments (actually, it is differential quadphase modulation in which the phase angle is determined with respect to the previous bit cell). This allows two input bits, which together have four states (00, 01, 10, and 11), to be coded into a single output bit cell. The modem input is 1200 bps but the output is only 600 bps. The 2400-bps modems that have become common combine amplitude modulation (AM) with phase modulation (QAM; quadrature amplitude modulation) to double the possible states yet again. The 2400-bps modem also sends just 600 bps over the telephone line. The high-speed modems that are now available use additional phase and amplitude states, together with adaptive line equalization and sometimes data compression, to provide rates of 9600 bps or more over good telephone lines (they automatically fall back to 2400 or even 1200 bps if the line is not so good).

This highlights the much misunderstood distinction between (data) bits per second and the baud rate. The rate at which the smallest signal elements (bits) flow through the transmission path is the baud rate. The rate at which the incoming data bits flow is, of course, the data bit rate in bits per second (bps). For systems such as straight NRZ coding in which one transmission bit represents one data bit, the baud rate and the data bit rate are numerically the same. In schemes that pack more than one data bit into each transmission bit, they are different.

Teletypewriter format

The serial format developed long ago for teletypewriters is worth looking at in some detail because it was the basis for later methods. It's essentially the format used by many modern serial devices including the PC serial ports.

The first TTYs used straight NRZ coding, like Fig. 3-7, but with characters (data) represented by a five-bit code. The problem of synchronization greatly bedeviled the first teletypewriters. The solution, apparently developed by E. E. Kleinschmidt early in this century, was to add synchronizing bits to the serial data. The key element was a start bit at the beginning of each character. Mechanical TTYs pause briefly after each character to allow time for the mechanism to reset and this idle period became a stop bit. (The slow mechanical decoding generally required more than one bit period to reset, and 1½ or 2 were commonly used.)

Because the added synchronizing bits "frame" the data, the total package—data plus sync—became known as a *frame*. An example of a frame using eight-bit data is shown in Fig. 3-11. Here is how it works. The stop bit and the idle state are at mark (high) level. The start bit, however, is at space (low) level. The transition from mark to space alerts the receiver that data bits follow. The receiver, which has been designed or programmed for N-bit transmissions, then accepts the next N bits as data. The $N + 1$ bit is the stop bit, which will be at mark level. If it is not a mark then it can't be a stop bit, and in that case the receiver knows that something has gone wrong and discards the entire frame (perhaps reporting a framing error). The receiver then starts over at the next high-to-low transition; that is, the next start bit.

This method synchronizes frames, not bits. But it is much easier to maintain reasonable clock synchronization over the few bits of a frame than over the thousands of bits of an entire message transmission. Synchronizing the receiver to the 60-Hz mains was adequate for the slower rates used by the old mechan-

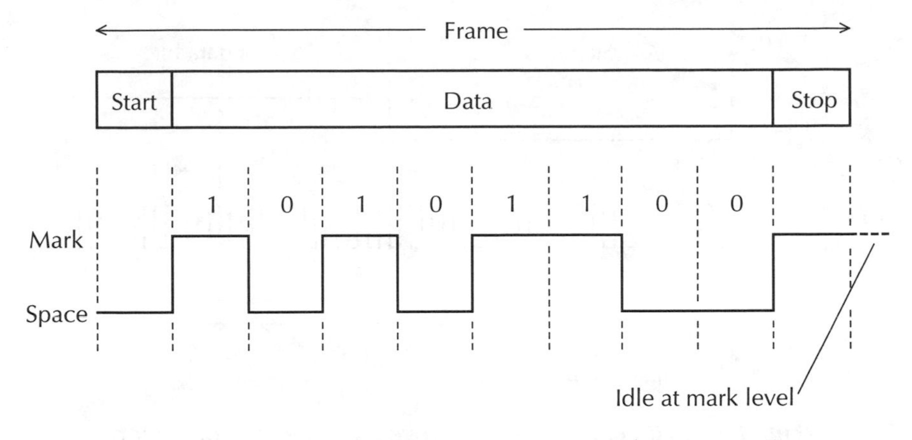

Fig. 3-11 *The TTY (start-stop NRZ) frame.*

ical TTYs. Modern crystal-controlled electronic systems easily achieve clock accuracies to a hundredth of a percent and readily handle serial streams at several tens of kilobits per second.

Although this transmission format is commonly called asynchronous serial, that designation is misleading. It is asynchronous *only* in the sense that frames need not follow each other at a regular rate. For the sake of accuracy, and in recognition of its historical origins in teletypewriter technology, I will call it TTY frame coding. Though simple, it has proved to be remarkably robust.

Most modern systems, including the PC serial port, use a more accurate method of achieving frame synchronization. The receiver clock is made some multiple of the bit rate, usually sixteen times as illustrated in Fig. 3-12. When the receiver detects a mark-to-space transition it begins counting clocks. If the line is at space for eight clocks, a valid start bit is assumed. The receiver is now positioned at the center of the start bit cell. The receiver then samples the input every sixteen clocks thereafter until all the data bits and the stop bit (or bits) have been processed. In this way, the receiver is phase-locked to the center of the bit cells with an accuracy of $\pm 1/16$ bit as it begins assembling the data byte, thus providing maximum leeway for any clock drift. This is a simple form of burst synchronization, and if you are familiar with color television you might notice its similarity to color subcarrier synchronization.

Circuits to generate the TTY frame can be assembled using discrete shift registers and gates, but nowadays it is most always done with a special-purpose chip—the universal asynchronous

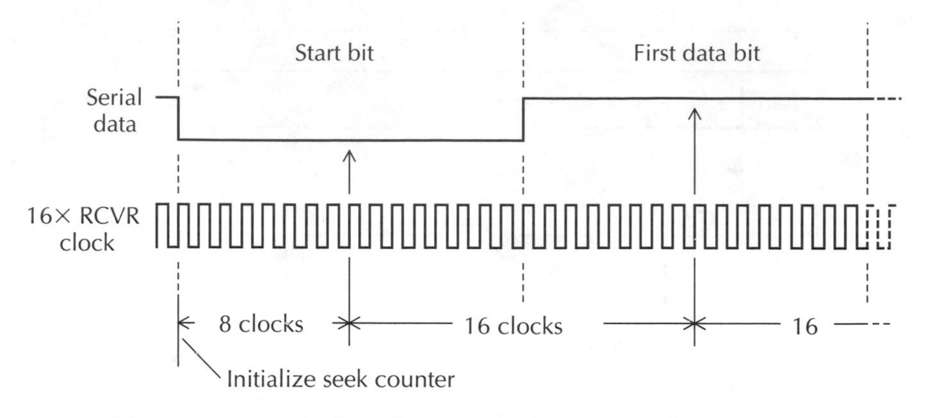

Fig. 3-12 *16× clocking for improved frame synchronization.*

receiver and transmitter (UART). Generating or decoding a TTY frame requires nothing more than reading or writing an I/O port. Chapters 4 and 6 discuss UARTs in greater detail.

There is, by the way, a difference between line coding and frame coding. The *line code* is the electrical format of the bits of the signal. The *frame code* is the logical format of a frame. A given frame coding can be transmitted with different line codings. The TTY frame shown in Fig. 3-11, for example, is in NRZ line coding, which is how it comes out of a PC serial port. If such a signal is sent through a modem, the frame coding remains the same (start-data-stop), but the line coding is changed considerably.

Synchronous transmission

Most big-time serial data transmission uses so-called synchronous techniques. As with asynchronous, the term synchronous is misleading. All serial transmission is fundamentally and necessarily synchronous—that is the only way to retrieve the data bits from the stream. Synchronous and asynchronous serial operations are so similar that it is quite practical to make UARTs that can be set for either mode, as in the Intel 8251A.

Synchronous systems differ from TTY transmission in three ways. The first and biggest difference is that the frame is usually a block of data rather than a single character. Second, it is common to use techniques that achieve receiver clock synchronization during as well as at the start of a frame. (One or another form of self-clocking line coding is often used.) Third, special characters are ordinarily sent during idle intervals to maintain receiver synchronization during this time also.

Because our concern with serial transmission is primarily with the TTY frame coding used by the PC serial port, synchronous methods are not discussed in detail. There are two that do merit a very brief summary, however; both have been offered by IBM as options for the PC.

Bisync A synchronous system that has been widely used in the past is IBM Bisync (BSC). Data is organized into blocks of bytes which are transmitted at a constant rate. Whenever there is no data, the transmitter automatically sends out sync characters (0011 0010 = 32 hex). The receiver establishes synchronization by waiting for two contiguous sync characters, denoting the start of a frame.

HDLC and SDLC At present, the most widely used synchronous formats are high-level data link control (HDLC), which has been standardized as ISO 3309, and the similar IBM synchronous data link control (SDLC). Data is transmitted as blocks which are framed by a special framing character. Unlike most previous methods, however, the basic data element is not the byte but the bit. How the bits in the data area of the block are used is left up to the individual application; most often, of course, they are grouped as eight-bit bytes or sixteen-bit words. The HDLC frame is discussed in greater detail later in this chapter.

The start and end of frames are demarcated by a unique character, the eight-bit sequence 0111 1110. A trick is used to ensure that no data ever has this pattern. Whenever the data has five sequential 1s, the transmitter automatically inserts a zero bit. The receiver finds the start of a frame by detecting six 1s in a row. It then switches over to frame mode during which it automatically discards the first zero bit after any run of five 1s.

Transmission standards and practices

The electrical characteristics of parallel transmission systems tend to be fairly simple. The Centronics interface, for example, is composed of ordinary unterminated TTL lines. High-performance parallel systems such as the GPIB pay attention to line termination (a topic that is pursued in chapter 7), cross talk, and similar matters, but otherwise are usually similar to TTL circuits. The situation is different with serial transmission systems. Quite a number of different transmission standards and methods exist. Let's take a look at a few of the most common ones.

Current loops

The early telegraph and TTY systems used current-loop technology, a practice that survived until dot matrix printers began to replace the electromechanical TTY. In a current loop the signal is a current, not a voltage. If several devices are connected to the same line they must be hooked up in series.

In classical TTY practice the nominal line current was 60 mA, fairly hefty by modern standards. The 1 or mark level was defined as current flowing and the 0 or space level was current off. The idle state was defined as mark level because this allowed a broken circuit to be detected immediately. If the line remained at space level for longer than a frame, the break condition (well, what else would you call it?) existed. Before long operators began forcing this condition on purpose as a special "attention" signal, a use it has had ever since.

Current-loop technology has been around a long time, and it is still very much alive and well. Nowadays we use optoisolators rather than relays and typical line currents of 20 mA or 5 mA. The MIDIs widely used with synthesizers and other electronic musical instruments, for example, uses a 5-mA current loop. You would do well to keep the current-loop method in mind when designing data acquisition or transmission setups. It is simple, reliable, inexpensive, and transmits dc (and thus straight NRZ data). It also provides full ground isolation—very important in some applications. A further often overlooked advantage is that it is inherently low impedance and can provide both very wide transmission bandwidth and very good noise immunity.

The RS-232 interface

The most widely used serial interface is the one set out in the Electronic Industries Association (EIA) recommended standard RS-232 (the current revision level is D). It's what most people have in mind when they use the word "serial." It should be emphasized that RS-232 is not concerned with TTY frame coding, but with modem control. Such "synchronous" systems as IBM's BSC and SDLC can and do use RS-232. It is, as its title states, a standard for the interface between data communication equipment (DCE)—in a word, modems—and data terminal equipment (DTE)—computers, printers, and everything else.

The RS-232 standard defines a number of circuits that connect a DCE (modem) to a DTE device. Table 3-1 lists the principal RS-232 circuits by code letters and name, with the more common abbreviations in parentheses. There are a number of other circuits that are less often used and are absent from PC serial ports. The signal flow is shown in Fig. 3-13. (The connector pins for these lines are shown in Fig. 4-2 in the next chapter.)

Table 3-1 RS-232 circuits.

Circuit	Function
Group A—grounding	
AA	Protective ground (chassis)
AB	Signal ground
Group B—data	
BA	Transmitted data (TD or TX)
BB	Received data (RD or RX)
Group C—control	
CA	Request to send (RTS)
CB	Clear to send (CTS)
CC	Data set ready (DSR)
CD	Data terminal ready (DTR)
CE	Ring indicator (RI)
CF	Received line signal detect (RLSD or CD)

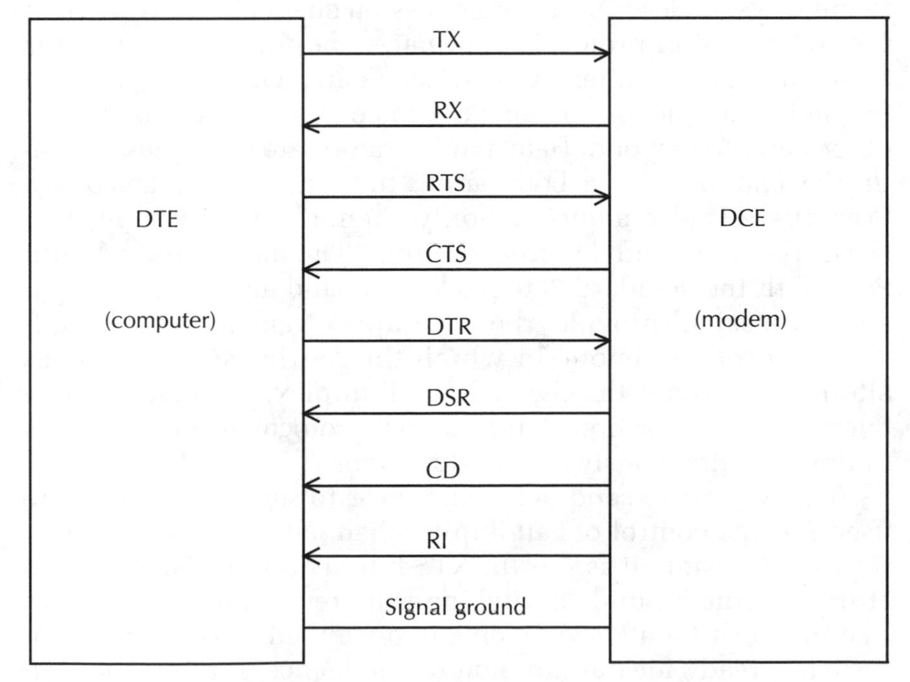

***Fig. 3-13** Major circuits of the RS-232 interface.*

The RS-232 circuits are divided into three groups. Only group B actually carries data. Data transmission requires only three wires: BA (transmit), BB (receive), and AB (ground). The circuits in group C are concerned with modem control. More exactly, they are for the control of what are now called "dumb" modems, which is all there was in 1960 when RS-232 was originally written. Modern "smart" modems use commands in the data stream for control.

Ready control The DCE device (modem) asserts data set ready (DSR) when it is connected to the line ("off-hook"), is not dialing, and is ready for transmissions. The DTE device (computer) asserts data terminal ready (DTR) when it is ready for business. Dropping DTR tells the modem to conclude transmission and to go "on-hook" (that is hang, up the phone).

Channel control The received line signal detect (RLSD), also known as carrier detect (CD), is asserted by the modem when it has detected a valid carrier of acceptable quality. Smart modems do this by emitting the CONNECT keyword. The ring indicator (RI) line is, as the name suggests, activated by the telephone line ring signal and indicates an incoming call.

Line control Let's take a moment to talk about half-duplex and full-duplex transmission. These terms are used in two different ways: to describe the transmission channel and to describe the transmission protocol or format. A *half-duplex channel* is one which is used alternately by each party. Only one party can transmit at a time; the channel has to be "turned around" for the other party to respond. Ham and CB radios are examples. A *full-duplex channel* allows both parties to transmit simultaneously. (Yes, there is also a one-way-only channel: simplex transmission, like radio and TV broadcasting.) The modems used with PCs (both the standard 300-1200-2400 baud and the new high-speed types) all provide true full-duplex transmission. A half-duplex *protocol* is one in which the parties send messages alternately, even if the channel is full-duplex, as in an ordinary telephone conversation. A full-duplex protocol allows messages to flow simultaneously in both directions.

The request to send (RTS) and clear to send (CTS) lines are used for line control of half-duplex channels. When one system wants to transmit, it asserts the RTS line. This tells the modem to "turn the line around," switching from receive to transmit. It's like the push-to-talk switch on a two-way radio. When the DCE device is ready for transmission to commence, it raises the CTS line, telling the DTE device to go ahead.

The RTS and CTS lines have very limited functions when the channel is full-duplex. The RTS just tells the modem to suspend transmission. CTS essentially duplicates DSR. Regrettably, and confusingly, some modems have a so-called half-duplex command that is really just a local echo control.

It is clear that the RS-232 control signals have significance only for modem control, and even then primarily with dumb modems. (An exception is the ability to use DTR to force a smart modem to go on-hook and into its command mode.) They are not necessary for data acquisition and transmission applications and they will not be used in the applications discussed in this book. Nevertheless, they are there and available as extra on-off control lines should you need to use them. As such they can be quite useful in some cases. Also, some serial devices such as printers and plotters use one or another of these lines to effect RDY/BSY hardware flow control.

The RS-232 standard covers the DTE-to-DCE connection. Because everything except modems are DTE, the question often arises as to how to connect one DTE device to another DTE device. This is easy if the principles illustrated in Fig. 3-14 are kept in mind. The situation envisioned by RS-232 is shown in configuration A. Modems are supposed to be transparent to data, so as far as the data is concerned the connection in configuration B is indistinguishable from the one in configuration A. Because there are no actual modems involved, the control lines can be ignored, arriving at the interconnection shown in configuration C. The TX line of one DTE device goes to the RX line of the other, and vice versa. Simple!

It is possible that you might run across software that requires the RS-232 control signals. Such software can be accommodated (that is, fooled into thinking it's connected to a modem) by a "null modem" cable. The TX and RX lines are as in Fig. 3-14c. In addition, RTS is simply strapped to CTS on the same connector; and likewise DSR and CD are connected to DTR.

Electrical characteristics of some common serial interfaces

Current loop The current loop is a simple on-off circuit and the principal specification is the nominal loop current. Serial current loops nowadays almost always use LED-photodiode optical couplers, and the more or less standard loop current levels are 20 mA and 5 mA. A complete specification for a current loop must specify more than a nominal line current, however. Every

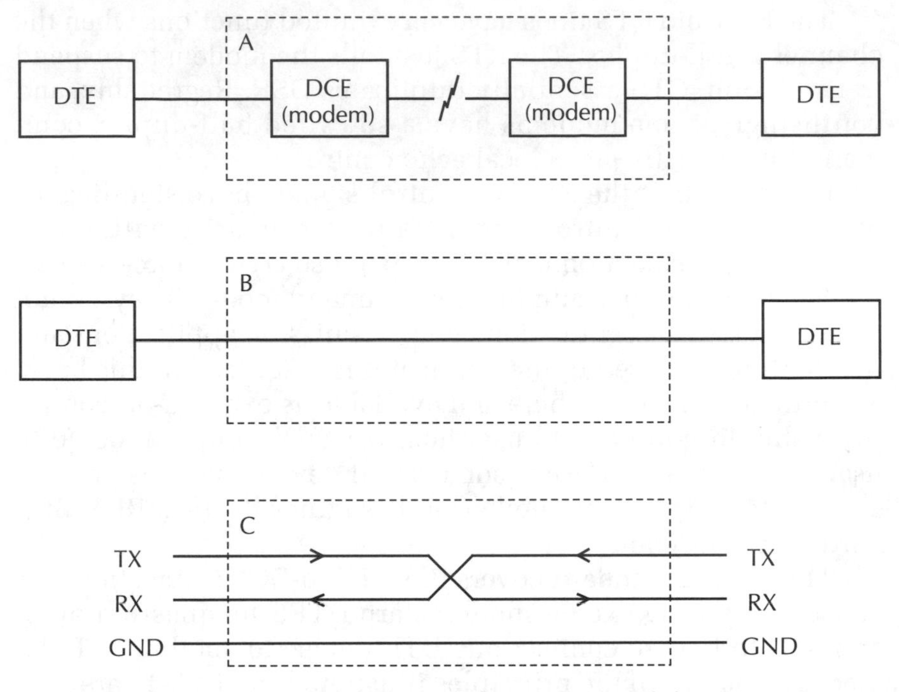

Fig. 3-14 *Interconnection of two DTE serial devices.*

loop has some resistance, which might become appreciable over long distances. A complete specification should include a maximum loop resistance.

The source supplying the current must have a maximum output voltage (compliance) sufficient to drive the rated loop current through the maximum loop resistance. Nowadays current loops are often implemented with open-collector drivers and series resistors to the 5-V logic supply, and it is usually assumed that the loop resistance will be kept small enough so that the loop voltage drop will not be appreciably greater than the typical LED drop of 1.2 V to 1.6 V. A better current-loop driver is described in chapter 7.

Standard RS-232 The threshold levels specified in RS-232 are shown in Fig. 3-15. Note that mark, which is the idle state for the standard TTY frame, is a negative voltage and space is a positive voltage. The RS-232 uses unbalanced or single-ended transmission, in which the signal is represented as a voltage with respect to ground. The input resistance of the receiver is specified as 3 to 7 kΩ, and the output impedance of the driver is typically 300 Ω. The maximum slewing rate of the signal is specified as 30 V/μs. At this rate the typical rise time will be 1 to 2 μs,

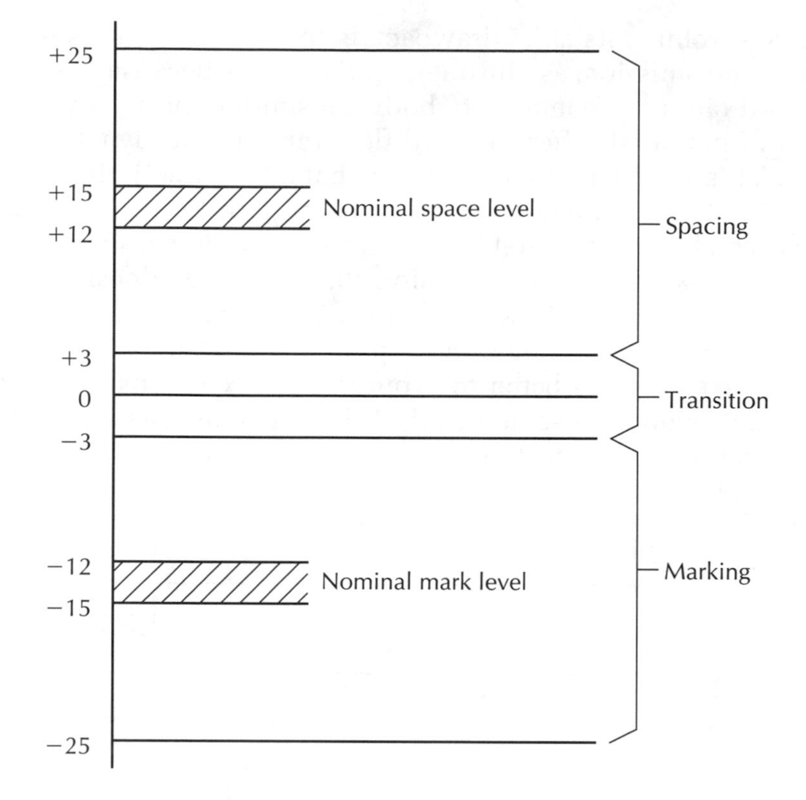

Fig. 3-15 *RS-232 interface signal levels.*

which is adequate for data rates up to 100 kbaud or so. The standard specifies a maximum cable capacitance of 2500 pF. The standard does not specify a particular connector, but the twenty-five-pin Canon series D is widely used.

The recommendation of a maximum line length of 50 feet is based on a cable capacitance of 50 pF per foot. That is rather conservative; coaxial and high-quality audio cable is typically closer to 25 pF per foot, giving a maximum length of 100 feet. Most RS-232 drivers can maintain adequate rise times with loads greater than 2500 pF, so if good quality cables are used even greater distances are possible. These distances are based on transmission at the maximum rate; at lower rates, such as 9600 baud, significantly greater transmission distances are possible.

The European CCITT standard V.24 is the same as RS-232. Military standard MIL-STD-188C is similar, except that the minimum receiver input resistance is 6 kΩ and the slew rate is controlled to be 5 to 15 percent of the bit interval.

RS-422 and RS-423 The RS-232 interface is simple, economical, and robust. Its chief drawback is the fact that it uses unbalanced transmission, as illustrated in Fig. 3-16a. Because the signal ground circuit is common to both transmitter and receiver, any ground potential difference will flow through the signal ground line. This produces a voltage drop that adds directly to the signal. Another problem is that the signal path forms a loop and noise can be induced in it by transformer action (because it's primarily magnetic induction, using shielded cable doesn't help). The relatively high signal levels in RS-232, $>\pm3$ V and typically ±12 V or so, help reduce these problems, but for high-performance systems it is better to avoid them. Two extensions of RS-232 that overcome some or all of these limitations are widely used in critical applications.

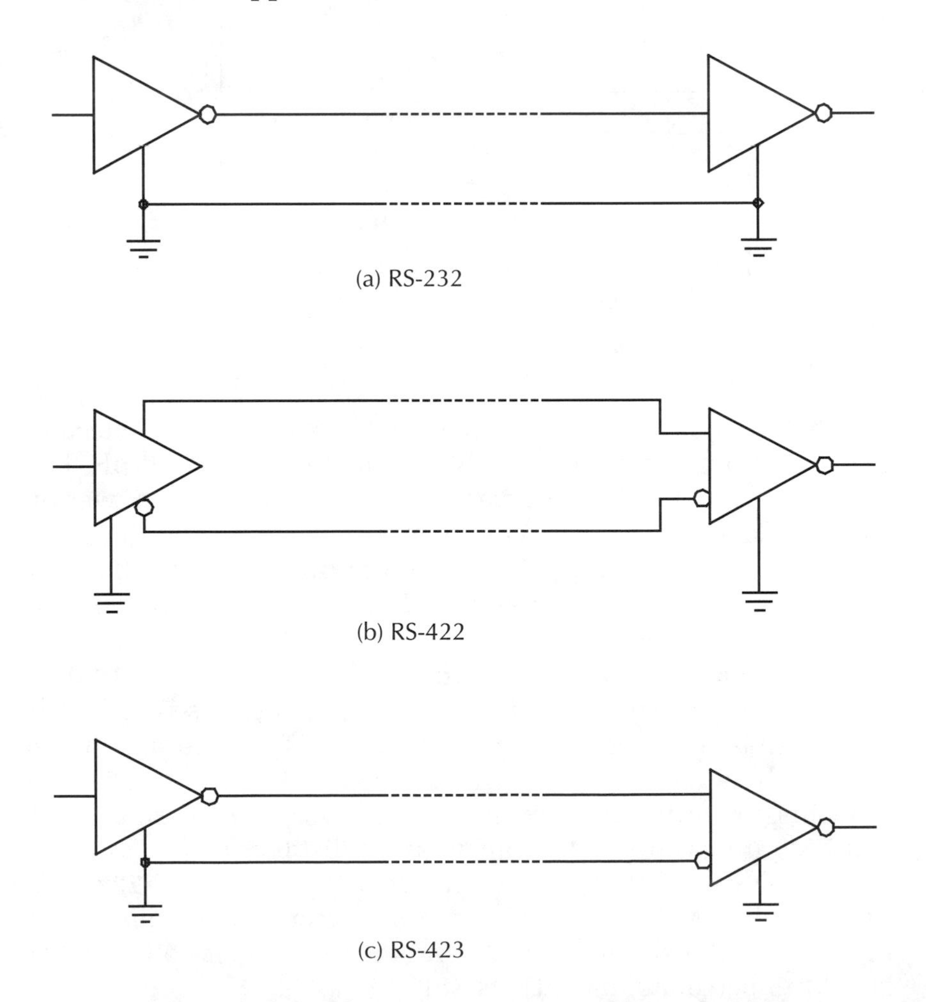

Fig. 3-16 *The most common EIA standard serial transmission configurations.*

The RS-422 standard uses a fully balanced transmission path, Fig. 3-16b. The receiver responds only to the differential signal on the pair of signal lines. Ground noise and induced signals appear in common mode at the receiver and are suppressed. The maximum voltage levels are ±6 V, and the typical receiver differential sensitivity is ±200 mV. Optional line termination is included in the specification, and drivers must have sufficient output current to drive a terminated 100-Ω line.

The RS-423 standard shown in Fig. 3-16c is a kind of halfway method. The same unbalanced driver as in RS-232 is used. The receiver, however, is balanced and its differential input is connected to the signal line and the driver ground reference. This provides increased noise immunity, but not nearly as much as RS-422.

Transmission lines By far the best performance is obtained by using properly terminated transmission lines, just as in radio frequency (RF) and wideband analog systems. One of the most widely used is the system originally used in the IBM 360 mainframe system. Terminated 93-Ω coaxial cable is used, driven with essentially TTL-level signals. Special drivers and receivers are used; ordinary TTL parts are not satisfactory for driving transmission lines. This method is discussed in greater detail in chapter 7.

Block transmission

Although usually associated with communication networks, the principles of packet transmission are not without interest for data acquisition systems. For one thing, data transmission is becoming more and more common even in modest applications, such as sharing data over a local area network (LAN) or just using a modem to send or receive data from someone else. Even more significant is the contribution that structure makes toward organizing data and keeping it straight. This is no small matter; many of us find ourselves dealing with increasingly large amounts of data. Structuring data as blocks or "packets" can help in keeping it organized and make its manipulation and analysis more efficient.

Let's look first at packets from the standpoint of transmission and data interchange. A *packet* is a block or array of data bytes or words, usually with some additional data applicable to the data block. From a programming standpoint, it is much like a Pascal record or a C structure. However, packets are included in frames which also have items concerned with such things as the size, nature, and integrity of the packet itself.

In recent years an internationally recognized format for packet transmission has been established as part of CCITT standard X.25. It, or very similar arrangements, are used in the HDLC, SDLC, and several other protocols. The general layout of an X.25 data ("information" or I) frame is shown in Fig. 3-17. The lengths of items in an X.25 frame are specified in bits rather than bytes and need not be in multiples of eight bits.

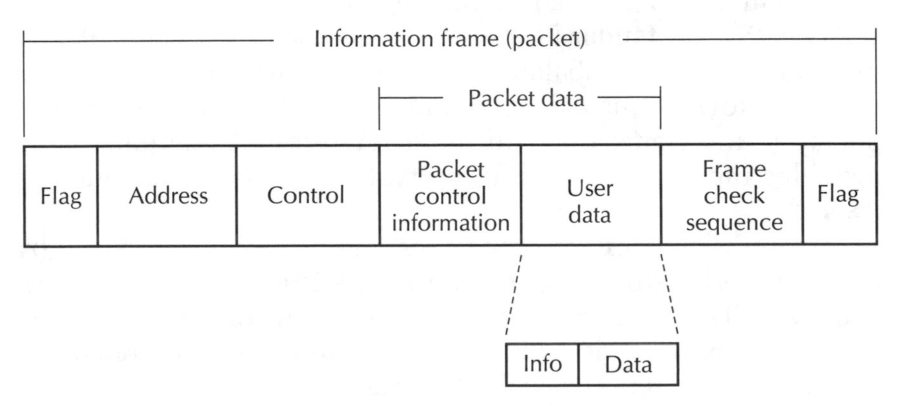

Fig. 3-17 *The CCITT X.25 packet.*

While most of us probably will not deal directly with X.25 packets, they are a useful conceptual model and are worth looking at from that standpoint. The flags are used to mark the beginning and end of the frame and need not concern us. The frame itself has several distinct parts. There is an address field indicating destination. The control field indicates the type of packet and is used for certain aspects of link control. The information field contains the actual data, perhaps with its own header containing information about the data, such as its length. A frame check sequence (FCS) field holds a check value for error detection. What is noteworthy is the clear demarcation of functional levels. There is the data itself and (optionally) data about the data; these together constitute the information. The remaining levels concern getting the information to the right place. This kind of insulation of logical entities is useful, and becomes increasingly useful as the complexity and variety of the data system increases.

Using packets in data acquisition

Consider the matter of one's own data files. The concept of organization used in packet systems can be helpful. Fundamentally packets are a useful way of organizing data items that tend naturally to be grouped, such as in a scientific experiment in which something is stimulated at intervals and each stimulation produces a set of data. For that matter, even text can be organized in such a way, for instance as paragraphs. One might want to regard packets in a data file as "subfiles" or something of the sort.

The concept of a packet has much in common with the "records" or "structures" of high-level languages. Indeed, a packet can be regarded as an object or entity, somewhat in the manner of object-oriented programming. The main difference is that packets presuppose many instances of the structure and, furthermore, that individual records are related to one another in some overall fashion(such as being sequential in time). Seen from this point of view, packets are more complex than data structures but less elaborate than databases. Data organized into packets lends itself very nicely to being indexed. This becomes a powerful tool for accessing as well as manipulating and analyzing the data. A simple application might require no more than a list; more complex instances would use a more powerful indexing method such as a B-tree or hash tables.

Some might consider that these ideas are well and good as theory, but excessive for everyday, real-world use. But in truth there are few things so practical as a really good overall data structure. Getting into the habit of ferreting out and looking at the structures implicit in the data can pay off handsomely. The structure itself, growing as it does out of relationships inherent in the data, can shed further light on the data. Everyone knows how useful it is to plot a set of data points. The information shown in the graph was there all along, but plotting it makes it easy to see.

Figure 3-18 presents the skeleton of a simple generic data packet format to illustrate the idea, and you can adapt it to your particular needs. The overall packet structure is given by *FrameRec,* some of whose fields are other records. Building structures as records of other records makes it easy to modify things later on or to change the structure for use in a different application.

The information in the packet is divided into a control record and an information record. The control record contains information about the packet itself. The sequence number *(SeqNr)* keeps track of the order of packets. This means that the packets do not necessarily need to be stored sequentially on disk or in memory, giving you an extra degree of freedom. Less obvious, perhaps, is the fact that reordering or shuffling packets (for ranking, alpha-

Fig. 3-18 *Structure for a simple data packet.*

```
Packet format: | Seq | Len | Data | FCS |

TYPE
   ControlRec = record                                       { control group }
                   SeqNr,
                    Len : word;
                end;
   InfoRec = record                                             { data group }
                { any information about the data can go here }
                Dta : array [1..MaxDtaSize] of byte;
             end;
   FrameRec = record                          { assembles groups into packet }
                 Control : ControlRec;
                 Info : InfoRec;
                 FCS : word;
              end;
```

betizing, or the like) might entail no more than changing the sequence numbers. No physical moving of records is required. The other control entry is the length of the data record in bytes. It is helpful to store the length explicitly because it can greatly speed up processing of the block; the routine that processes the block need not hunt for the end of the data.

The structure of the *information record* will depend on the application. In some cases it could be an array of bytes or a string of text. Other applications might require a more complicated explicit structure. It often happens that there are certain items that refer to a block of data as a whole; these can be included as a header record in the information record also. For example, the data might be an array representing digitized voltage for a short time following a stimulus. The data header could include information on the kind, intensity, and time of the stimulus, and so forth. The size of the information record depends on the application. When the choice is relatively arbitrary, a size of 1024 to 4096 bytes works well.

FrameRec also includes frame check sequence (FCS), a sixteen-bit field intended for an error-detection code. This field is discussed in greater detail later in this chapter.

Local area networks

Local area networks (LANs) are becoming increasingly common, and are quite valuable for handling and exchanging data in scientific, engineering, and business applications. Most of the time the data is shared or exchanged through common access to files on a file server or by moving data across the network in the form of files. What actually flows through the cables connecting the network is nothing other than data packets of various kinds. Many networks allow users to send data over the network as packets rather than in the form of conventional files. This is often referred to as "peer-to-peer" communication because it need not go through the central file server. It can be quite useful and fairly efficient.

The popular Novell Netware, for example, provides internetwork packet exchange (IPX) for direct packet transfer. Later versions also provide sequenced packet exchange (SPX) which provides verification that packets are received correctly; IPX uses "blind" transmission with no means for verifying reception. Each packet is a header with various connection information and the actual block of data to be sent. The user's data block can be as large as 546 bytes under IPX or 534 bytes under SPX. Packets can be sent to a particular destination (or "socket" in LAN jargon) or broadcast over the network.

Organizing data in the form of packets naturally fits in well with this scheme. Using IPX or SPX ordinarily requires calls directly to the Netware kernel. The process is straightforward, but there are a number of details that have to be seen to. Those who are interested in this very effective use of network resources should refer to a reference manual such as Rose (1990).

Transmission protocols

Data transfer involves two things: the way the data is formatted, which is what we have been looking at; and, secondly, some orderly procedure—protocol—for sending and receiving it. The latter will be familiar to many in the form of file transfer protocols such as Xmodem or Kermit. In point of fact, though, all data transfer has both elements. Sometimes the process is so simple and familiar that it is not thought of in these terms, such as in transferring data between memory and disk. The data format in that case is an array of bytes and the protocol is issuing a read or write call and checking for any errors afterward. Flow control of data I/O is an implicit protocol.

There are a great many formal protocols in use in data transmission systems. Some are relatively complex, as in GPIB systems or HDLC packet communication. Others are more simple. Let's take a look at one of the simpler file transfer protocols, the popular Xmodem. There are three reasons for doing so. One is that it illustrates the concept. The second is that you might sometimes find it useful to use Xmodem directly rather than through a communications program for transferring data—cutting out the middleman as it were. The third is that it shows how not to design a protocol.

The Xmodem packet is shown in Fig. 3-19. It consists of the ASCII start-of-header symbol, Ctrl–A, a sequence number, the negation or one's complement of the sequence number, a fixed 128-byte data block, and a frame check. The frame check in the original Xmodem is a single byte and is the sum of all the 128 bytes in the data array, modulo 256. A later modification, Xmodem-CRC, uses a sixteen-bit cyclic check code over the data block. A full eight-bit transmission path is required.

An Xmodem transfer proceeds as follows:

1. The transmitter awaits a start-up signal from the receiver. The receiver sends the start command. For the original checksum version, the start character is ASCII NAK (^U, 15h). For Xmodem-CRC it is a capital C (43h).
2. The transmitter assembles the first packet, for which the sequence number is 1, and sends it.
3. The receiver examines the first incoming byte.

   If it is SOH (^A, 01h) the receiver knows that a packet is being sent. The receiver assembles the packet and checks it. If the check code shows it is correct, the receiver responds with ACK (^F 06h); if there is an error, the receiver responds with NAK and the transmitter retransmits the packet. (Error detection is discussed in detail in the next section.)

   If it is EOT (^D, 04h), indicating end of transmission, the receiver ends the session.

   If it is CAN (^X, 18h), the abort command, the receiver signals that the session is aborted and stops receiving.
4. During the session the transmitter continues to send packets and the receiver responds ACK if good or NAK if bad, until the last packet is sent or the session is aborted.

Fig. 3-19 *The Xmodem packet.*

```
XMpacket = record
        SOH,                                  { always ^A (01 hex) }
        SeqNr,                                        { 1 to 255 }
        NotSeqNr : byte;                      { negation of SeqNr }
        Dta : array [1..128] of byte;               { fixed size }
        Chk : byte; { or word }                    { frame check }
    end;
```

The Xmodem protocol works reasonably well and is very widely used in the PC community. However, it has a number of weaknesses. The most serious failing is its fixed block size. If the data is not an exact multiple of 128 bytes, the last block will be only partly filled. There is no way to specify how many data bytes in a block are valid. A second problem is that it has no provision for any information about the data, such as a file name or file size. This information could be put into the first data block, of course, but because there is no standard format for so doing, that ploy would be useless for transferring files with other users. The remaining drawbacks are less serious. The inclusion of the negation of the sequence number is presumably to permit error checking, but it is a weak form of error detection. A wiser design would have included the sequence number in the check code domain. It is also a single-shot protocol; there is no standard way to send multiple files. Despite these limitations, it is not without value for general data transmission, mainly because it is so widely used.

Error detection

Most people think of error detection in connection with storage devices and especially with telecommunication systems. Few would think of adding explicit error detection to a data acquisition system and to their own data files. Getting an instrumentation setup to yield good quality data is not always easy, but many people feel that once it's "in the computer" it's safe to relax. It is true that modern disk and tape storage is remarkably reliable, but it isn't perfect. Considering the effort you have probably gone to in acquiring the data, it just makes good sense to have some means of catching any errors that might creep in.

Error control encompasses two related but distinct areas: error detection and error correction. The latter are methods for recovering correct data after it has been damaged. They tend to be complex, generally must be optimized for particular situations, and have limits as to the amount of damage that can be undone. The error correction system in audio CDs, for instance, can handle dropouts because of dust and wear, but not large smudges. Error detection methods make no attempt at correction, and consequently are generally simpler and also have greater generality. I will look only at error detection.

Two of the most important criteria for evaluating the quality of an error detection method is its sensitivity, or how good it is at catching errors, and its efficiency, or the degree to which it slows things down or consumes transmission or memory space. These criteria tend, on the whole, to involve a trade-off. For example, comparing a copy with a known good original, item by item, has perfect sensitivity but is extremely inefficient.

The error detection methods considered here are those that associate some kind of *index* with the data which allows a determination (with greater or lesser sensitivity and efficiency) of whether or not errors have occurred. Very often these indices are embedded in the data stream itself, as in the case of check bytes appended to a block of data. Methods that append indices for error detection are often called *redundancy codes.*

Parity checks

One of the simplest error detection methods is the *parity check.* The index has an extra bit, the parity bit, which is added to a group of bits. The value of the parity bit, 0 or 1, is based on the number of 1s in the group. For odd parity, the value of the parity bit is such that it makes the overall sum odd; for even parity, the sum is even. A parity bit is often included in the TTY frame code format in computer systems. This used to be almost universal, but the trend is away from it. For ordinary modem communications, the format "eight bits, no parity" is becoming very common.

Parity is not very sensitive. A parity check will catch all single-bit errors and all errors that affect an odd number of bits. On the other hand, it cannot detect any errors involving an even number of bits. In practice, most errors are "burst" errors which involve several bits (most often, several bytes); on average, parity will catch only 50 percent of the errors. To add insult to injury, parity is not very efficient either. In the common seven-bit plus parity format, the data rate is reduced to 87.5 percent of what it would be without parity. (For eight bits with parity versus eight bits without parity, the rate is reduced to 88.8 percent.)

In fairness, there are some instances in which parity is useful. The PC uses parity to check memory (physically, PC memory is nine bits wide). Because individual memory chips are one bit wide, all single chip failures will indeed be caught.

Checksums

For a given error index size, efficiency goes up as more bits are included in the error computation. To take advantage of this fact, *block* methods are used. The basic idea of a block check is to use some mathematical procedure to derive an index for the block.

An early application of error checking to entire blocks was *longitudinal parity*. Think of a group of bytes stacked on top of one another. The bits in each byte form a row; the *N*th bit of the bytes line up in columns. Longitudinal parity is a parity check on such columns. The method grew out of tape storage systems. Standard computer tape is written a byte at a time, with a separate track for each bit. Longitudinal parity requires no more than a running sum on each channel. It is significantly more sensitive than ordinary parity, but inefficient to compute if the data is available only as bytes rather than separate bits.

Another method that has been around for a very long time is the *checksum*. (Since ancient times, in fact; manuscript copyists hit on the idea of treating letters as numbers and adding them up.) The checksum of a block of data is just the sum of all the bytes, usually modulo 256 or 65536 to limit the size of the index to eight or sixteen bits. The method is fast and very simple, but not very sensitive. Furthermore, the sensitivity decreases drastically as block size increases. The method is completely blind to zero bytes and to any reordering of the sequence of bytes in a block. For small blocks (not larger than, say, 96 or 128 bytes) the method is not too bad.

An ingenious modified checksum algorithm that is said to approach the excellent sensitivity of the cyclic codes described is presented by J. G. Fletcher (1982). Its sensitivity is greatest for block sizes of 255 bytes and smaller.

Cyclic codes

A powerful and increasingly common method of error detection is the use of cyclic or polynomial codes, particularly the cyclic redundancy check (CRC). An excellent and relatively simple summary of the theoretical basis for such codes is in Peterson (1961).

Cyclic codes have acquired a certain aura of mystery, but the basic principles are not difficult. In a nutshell, the idea is that an integer N of length (number of digits) m divided by a suitable number G will have a remainder r for each such N. (The quotient is ignored.) The value of r is an index for each N. The divisor G is selected to maximize the number of different remainders over all possible values of N to make the indices "as unique as possible." For error detection let N be a block of data bytes that is treated as a single m-bit number, and use the remainder r as the error checking code.

Cyclic coding theory uses polynomial arithmetic. Our ordinary pencil-and-paper arithmetic is polynomial over the field of integers in base 10. Recall that you write numbers in "positional notation" in which a number is really the sum of powers of 10, each digit being a coefficient. You could write a number such as 31,405 as $3 \times 10^4 + 1 \times 10^3 + 4 \times 10^2 + 0 \times 10^1 + 5 \times 10^0$. In addition and subtraction you take sums of each power—that is, each column of figures—separately, adding to or borrowing from the adjacent column if necessary. Multiplication and long division are a little more complicated; they are actually algorithms that use shifting together with addition and subtraction. To evaluate, say, 123×789 you go a column at a time (leftward, from 10^0 to 10^2) and get an intermediate product, each time shifting the partial product one place to the left. Finally, you add the partial products to get the result.

Binary information can be written in polynomial form in the same way, as powers of 2. Because the coefficients are either 1 or 0, terms with zero coefficients are customarily omitted; for instance, 01011 is $2^3 + 2^1 + 1$. Frequently a dummy variable is used; for example, 01011 is written as $x^3 + x + 1$.

A polynomial can be expressed as the product of smaller polynomials, or factors. However, a polynomial can be factored into prime factors—factors divisible only by themselves and 1—in only one way. Cyclic codes for error detection treat all the bits of the block as the bits forming one huge polynomial, which is divided by a prime factor (the *generator polynomial*), and the remainder becomes the index code for the block. The arithmetic is done modulo 2 (which is provided by the XOR operator), so there are no carries nor borrows. Unique factorization holds in modulo 2 also. Because the only operations required are shifts and XORs, calculations can be done at very high speed directly in hardware.

By far the best way to see how it works is to sit down with pencil and paper, write out a number and a divisor in binary, and do the division by hand. Proceed as in ordinary long division, but ignore borrows and carries: that is, $0 + 0 = 0$; $0 + 1 = 1$; and $1 + 1 = 0$. (Subtraction is identical.) A tip: to avoid getting bogged down in long strings of binary digits, make up a short CRC, such as a four-bit remainder with a divisor (generator polynomial) of, say, 1 0011. If you do want to use the full sixteen-bit CCITT polynomial, it is 1 0001 0000 0010 0001.

Practical cyclic error codes modify the process slightly. In an N-bit CRC calculation N zero bits are appended to the data before the division to ensure that all data bits will affect the remainder (or, equivalently, the data block is shifted left N places). A rearrangement of the division algorithm or shift register configuration can give the same result without the need for adding zero bits or an explicit left shift.

The sensitivity of a CRC depends on the choice of generator polynomial. The one most widely used for sixteen-bit CRCs is the CCITT polynomial, $X^{16} + X^{12} + X^{5} + 1$. (Yes, this is a seventeen-bit number; the generator polynomial for an N-bit CRC is of degree $N + 1$. The size of a CRC refers to the number of bits in the remainder.) The CCITT CRC has these excellent sensitivities:

- one-bit errors: 100 percent
- two-bit errors separated by fewer than 2^{16} bits (8192 bytes): 100 percent
- burst (adjacent bit) errors of up to sixteen bits: 100 percent
- all other errors: $1 - 2^{-16} = 99.998$ percent

Practical CRC algorithms

Harnessing the power of cyclic error detection codes for keeping watch on your data calls for a simple and rapid way to compute CRCs. Ideally you would have a simple library routine that would receive the data and return a CRC. That is what is offered in the CRC Toolkit shown in Fig. 3-20.

The CRC Toolkit, and the other toolkits in this book, illustrate the advantages of modular programming and reusable code. These ideas are important enough to warrant some comment. Both are rooted in ideas that go back to the early days of computing, when frequently used sections of code were assembled into libraries that could be linked into various application programs as required. Languages like C and FORTRAN that generate modules in the form of ordinary object language (.OBJ) files clearly show this heritage. In several more recent languages the notion of a program module is made an explicit part of the language. One of the best-known examples is the Modula family of languages originated by Niklaus Wirth, who also designed Pascal. Beginning with version 4, Borland extended Turbo Pascal to include modules, called units, much like those in Modula-2. Among other things, explicit modules differ from older object language modules by containing more rigorous variable typing and module interdependency information.

The most important feature of modules is the ability to control access to variables and code. Only those items that are declared as

Fig. 3-20 *The CRC toolkit unit.*

```
{ ---------------------------------------------------------------------------- }
{                          CYCLIC REDUNDANCY CHECK MODULE                      }
{   computes forward CRC using CCITT polynomial, X(16) + X(12) + X(5) + 1     }
{ ---------------------------------------------------------------------------- }
(* CRC.PAS 1.1 ©1992 J H Johnson *)

UNIT
   CRC;

INTERFACE


procedure PresetCRC( N : word );                    { initialize to 0 or FFFF hex }
  { call PresetCRC before beginning a CRC calculation }

function CRCresult : word;
  { returns current CRC accumulator value }

procedure UpdateCRC( Dta : byte );
  { incorporates data byte in CRC, using direct calculation }

procedure FastUpdateCRC( Dta : byte );
  { incorporates data byte in CRC, using lookup table }

procedure BlockCRC( var DataBlock; Len : word );
  { updates CRC with a block of data using lookup table }
  {  DataBlock is array of bytes. Len is number of bytes in DataBlock to do. }
  {  CAUTION: Len not value-checked; must be <= size of DataBlock!           }

IMPLEMENTATION
{ ========================================================================== }

CONST
   GenPoly = $1021;                                { CCITT generator polynomial }

VAR
   PartRemTable : array [0..255] of word;
   Accum : word;                          { accumulator; holds current value of CRC }


{ ----- direct CRC computation ----- }

{ UpdateCRC incorporates data byte in CRC, using direct calculation }

procedure UpdateCRC( Dta : byte );
var
   K : byte;
begin
   for K := 1 to 8 do                              { loop to do the 8 bits in Dta }
      begin
         if ((Dta xor hi(Accum)) and $80) <> 0 then                { test hi bit }
            Accum := (Accum shl 1) xor GenPoly            { set: xor, shift Accum }
               else Accum := Accum shl 1;                   { clear: just shift }
         Dta := Dta shl 1;                                { now shift Dta left }
      end;
end;
```

Fig. 3-20 Continued.

```
{ ----- CRC calculation using lookup table of partial remainders ----- }

{ FillPartRemTable loads lookup table with upper-byte intermediate values }

procedure FillPartRemTable;
var
   K : byte;
begin
   for K := 0 to 255 do
      begin
         Accum := 0;
         UpdateCRC(K);
         PartRemTable[K] := Accum;
      end;
end;


{ UpdateCRC incorporates data byte in CRC, using lookup table              }
{   PartRem (partial remainder) value is data byte XOR high byte of Accum, }
{   and is index into lookup table. The 16-bit table value is then XORed with }
{   low byte of Accum.                                                     }

procedure FastUpdateCRC( Dta : byte );
begin
   Accum := (Accum shl 8) xor PartRemTable[hi(Accum) xor Dta];
end;


{ BlockCRC updates CRC with a block of data using lookup table                }
{    DataBlock is array of bytes. Len is number of bytes in DataBlock to do. }
{    CAUTION: Len not value-checked; must be <= size of DataBlock!            }

procedure BlockCRC( var DataBlock; Len : word );
type
   ByteArr = array [1..MaxInt] of byte;          { max block size 32767 bytes }
var
   K : word;
   DB : ByteArr absolute DataBlock;
begin
   for K := 1 to Len do
      Accum := (Accum shl 8) xor PartRemTable[hi(Accum) xor DB[K]];
end;
{ ----- general ----- }

procedure PresetCRC( N : word );
begin
   Accum := N;
end;

function CRCresult : word;
begin
   CRCresult := Accum;
end;


BEGIN
   FillPartRemTable;                                   { initialize table }
END.
```

public can be accessed by other modules; everything else is *private* to that module. This means that several modules can contain private variables or subroutines with identical names without confusion and without danger that the wrong one will be executed. Subroutines in a module can be revised and improved without affecting other modules or the main program. It also means that important variables can be safely hidden inside a module. For example, the procedure *PresetCRC* and the function *CRCresult* in the CRC Toolkit allow specific kinds of access to the internal accumulator variable but protect it from accidental change by any other part of a program.

In explicitly modular languages like Turbo Pascal and Modula, each module (unit) has an interface section that declares public variables and subroutines, and an implementation section that contains the private code. A very handy feature is the optional initialization section; any code in it will be executed automatically following start-up of the run-time kernel.

Modules are ideally suited for implementing reusable code. This is simply the sensible idea that the wheel need be invented only once. The key to reusable code is the design of subroutines that perform various commonly needed tasks and that are as general in nature as possible. An example in the CRC Toolkit is **BlockCRC,** a routine that calculates the CRC for a block or array of bytes of any type or size (within reason).

All of the routines in the CRC Toolkit differ from the basic CRC algorithm in one respect. The CRC algorithm operates on bits, and is so computed in hardware with shift registers and XOR gates. The data in a computer is stored and manipulated in bytes, however, and routines to do CRC calculations have to be written to operate on eight bits at a time.

The heart of the CRC Toolkit is the procedure **UpdateCRC,** which does modified polynomial division modulo 2, using a loop to do the eight bits of each byte. The CRC of a block of data is obtained by sending each byte of the block in succession to **UpdateCRC.** Intermediate results are accumulated in the static variable *Accum* which retains them between calls. After all the bytes in a block have been processed, *Accum* will hold the final CRC.

UpdateCRC has one drawback: it is painfully slow. On a typical 8-MHz XT machine it poked along at an average rate of about 340 μs/byte. A machine of AT-class or better would be faster, of course, but it would be nice if you could do better. And it turns out that you can.

It was noticed many years ago that software CRC calculations could be speeded up substantially with the use of an auxiliary look-up table. In a direct computation (such as **UpdateCRC**) the bits of the data byte immediately affect only the bits in the upper byte of the accumulator. This suggests a shortcut. The result on the upper byte of the accumulator is regarded as a partial value, and a table of all of the 256 possible single-byte partial values is precomputed. The modified CRC algorithm then proceeds as follows. You XOR the incoming data byte with the upper byte of the accumulator as before, but now you use that result as an index into the table of partial values. The partial value from the table is then added to (XORed with, since you are working modulo 2) the lower byte of the accumulator to complete the computation.

FastUpdateCRC implements a look-up table version of the CRC calculation. Note that it has none of the time-consuming looping and shifting of the direct computation. The difference is significant. On the same 8-MHz XT machine the average processing rate was 66 μs/byte—more than five times faster than direct computation.

The look-up table of partial remainders is created by **FillPartRemTable** which uses **UpdateCRC** with *Accum* reset to zero before each call. The table is located in the static data segment, but it could be allocated on the heap instead if desired. Because the table must be initialized before any fast CRC calculations—it needs to be initialized only once—it is called in the automatic initialization section of the unit. This could be changed to an explicit call by the main program if the brief time taken during run-time start-up is objectionable.

It is possible to do still better. When a loop is used to pass each byte in a block to a CRC routine, the repeated activation and deactivation of the CRC subroutine wastes time. **BlockCRC** applies the table-driven CRC algorithm directly to an entire block of data. The increase in speed, while not dramatic, is worthwhile. In a test using 256-byte blocks, **BlockCRC** ran about 40 percent faster than **FastUpdateCRC.** The average processing speed on an 8-MHz XT was 48 μs/byte, or just over 20 kb/s, over seven times faster than direct calculation.

BlockCRC is written to take any array of contiguous bytes as an argument. The exact length of the block in bytes must also be passed to the routine. As with all **var** parameters in Pascal, the parameter *DataBlock* is just a far pointer. Because no type is declared for *DataBlock*, the type *ByteArr* and the dummy variable *DB* are used to inform the compiler that the untyped parameter is to be treated like an array. No memory is allocated for these items. Less strongly typed languages such as C permit untyped parameters without such dummy typing.

Using the CRC

Applying cyclic error codes to data is straightforward. A CRC is calculated for the original data and is thereafter included with it as a check word— for example, by being written as an additional variable in a file, or included in a block transmission. (It is standard, by the way, to transmit the msb [upper byte] of the CRC first, followed by the lsb.) Verification of data after it is received or read tests it by again calculating the CRC of the data (a "local" CRC) and comparing it to the check word. If they are the same, all is well; if not, then not.

Basically obtaining the CRC for a block of data involves (1) clearing the accumulator to zero with **PresetCRC;** (2) sending the data to the CRC routine, either byte by byte to **FastUpdateCRC** or, better yet, in one whack to **BlockCRC;** and (3) fetching the result through **CRCresult**. This would indeed obtain the CRC as it is used in the popular Xmodem-CRC file transfer protocol. But virtually every other use of the CCITT CRC introduces a couple of modifications.

The standard use of the CCITT CRC initializes the accumulator to all ones (FFFFh), not to zero. The reason is that a zero accumulator would not detect any zero bytes that have somehow been added to the beginning of the data (admittedly an unlikely event, but in the world of computers strange things do happen). The second modification is that when the CRC is appended to a data block as a frame check (FCS), it is inverted (ones complement).

The pseudo code in Fig. 3-21 shows how CRC error detection can be applied to data packets. The packet format is the simple structure presented earlier in Fig. 3-18. In this example the packet is the variable *Fr* of type FrameRec. The convention of initializing the CRC accumulator to FFFFh is used, and the raw CRC is added to the packet in inverted form as an FCS. You'll note that the upper and lower bytes of the CRC returned by **CRCresult** are treated separately. The reason is that the order for sixteen-bit words in Intel microprocessors is lsb then msb. The code shown reverses this to duplicate the standard order for the FCS which is msb first, followed by lsb. The byte-swapping code is shown for clarity. A better way is to use the swap function in Turbo Pascal (or swab in C) which swaps msb and lsb, like this:

```
Frame.FCS := Swap( not CRCresult );
```

Fig. 3-21 *Example of how to use CRC on data.*

```
CRC error detection with the simple packet of Figure 3-18:

| seq nr | len | data | fcs |
|---- crc on these ----|

creating or sending:
   make a Frame Fr of type FrameRec
   put the data into Fr.Info.Dta array
   set Fr.Control.Seq and Fr.Control.Len to appropriate values
   compute CRC:
      PresetCRC( $FFFF );            { preset accumulator to FFFF hex
      BlockCRC( Fr.Control, SizeOf(ControlRec) );
      BlockCRC( Fr.Info, Fr.Control.Len );
      hi(Fr.FCS) := hi( not CRCresult);    { FCS is inverted CRC }
      lo(Fr.FCS) := lo( not CRCresult);
   send Frame record Fr to destination

verifying (reading or receiving):
   fetch a complete Frame record Fr, then do CRC:
      PresetCRC( $FFFF );
      BlockCRC( Fr.Control, SizeOf(ControlRec) );
      BlockCRC( Fr.Dta, Fr.Control.Len );
   now do the check word, first un-inverting it:
      FastUpdateCRC( not hi(Fr.FCS) );      { msb (upper byte) }
      FastUpdateCRC( not lo(Fr.FCS) );      { lsb (lower byte) }
      Ok := (CRCresult = 0);
   if Ok is TRUE then use the data
```

Notice that in processing the received packet the local CRC is not compared to the check word; rather, the check word is also included in the local CRC calculation (after being uninverted). If the local CRC and the check word are the same, the final local CRC result will be zero—a quick and handy test.

It is possible to boost the speed a wee bit by omitting the uninversions, in which case a good data block plus the inverted check word will always yield the "magic" number 1D0F hex (7439 decimal). Furthermore, if the check word immediately follows the data block, as it does in the example, you can incorporate its two bytes in the call to **BlockCRC.** In fact, you can do the entire record—control and info and FCS—with one call. The routine to verify the block would then look like this:

```
PresetCRC( $FFFF );
{ do data bytes plus check bytes }
BlockCRC( Fr.Control, SizeOf(ControlRec)+Fr.Control.Len+2 );
{ omit the two calls to FastUpdateCRC }
Ok := (CRCresult = $1D0F);
```

Xmodem-CRC

The special case of Xmodem-CRC deserves mention. The CCITT generator polynomial is used, but the accumulator is preset to zero and the CRC check word is not inverted. The CRC is calculated on the 128-byte data block only. An example of how the error detection is implemented in the Xmodem-CRC protocol is shown in Fig. 3-22, using a packet variable *Xm* of type XMpacket (see Fig. 3-19). Because the CRC directly follows the fixed-size data block, the receiver error checking simply calls BlockCRC to do all 130 bytes at once.

Fig. 3-22 *Example of how to implement the Xmodem-CRC protocol.*

```
sending:
     make packet Xm of type XMpacket
     set Xm.SOH, Xm.SeqNr, Xm.NotSeqNr
     load Xm.Dta
     PresetCRC( 0 );
     BlockCRC( Xm.Dta, 128 );
     Xm.Chk := Swap( CRCresult );

receiving (assumes first byte = SOH):
     get packet Xm
     check that Xm.SeqNr is expected packet number
     PresetCRC( 0 );
     BlockCRC( Xm.Dta, 130 );
     Ok := (CRCresult = 0);
```

Summary

The concern in this chapter has been with ways of getting data from one place to another, doing so with accuracy, and being able to detect when it is and isn't accurate.

Transmission systems fall into two broad categories: serial and parallel. These were discussed in some detail, with particular emphasis on the RS-232 and the Centronics standards which are the kinds of serial and parallel transmission used by the PC I/O ports. This is a background for the more specific and detailed explorations contained in the next two chapters of the serial and parallel I/O used in PCs.

There is more to transmission than the raw movement of data bytes. Most data has a logical structure whose elements are groups of related bytes. Some simple ways of organizing groups of bytes into records or packets and handling them as discrete logical objects or entities were considered. Finally ways in which the integrity of data can be tested—an important matter, because data can only too easily become corrupted in transmission—were discussed.

Using the serial port

This chapter delves into the standard PC serial port. It begins with the standard serial port hardware and how to program it. Next it takes up the implementation of I/O functions and how to incorporate them in programs. This is followed by a thorough treatment of interrupts, especially hardware interrupts, and how to write and use interrupt service routines. Although the discussion of I/O routines and interrupts is with reference to the serial port, the principles and techniques set forth are quite general and by no means limited to the serial port. Finally a library of general-purpose, high-performance software tools is developed for using the serial port.

One thing this chapter does not discuss is how to use the BIOS serial services or the DOS AUX device. They are so poorly implemented as to be practically useless. They use the modem control lines, yet provide no way to control them. There is no way to test whether a character has been received. If a character is not received, the serial input routine will sit and wait until one comes along or it times out. Worst of all, input can only be done by polling, which is quite apt to loose characters at faster rates (above 1200 baud) and during disk operations.

The PC serial port

In early PCs the serial ports were plug-in cards—the IBM asynchronous adapter or third-party equivalents. Later machines generally have at least one built-in serial port, usually directly on the motherboard. Fortunately, these arrangements are functionally identical in all important respects. The OPC and the XT provided up to two serial ports, commonly known as COM1 and COM2. The AT "supports" two more serial ports but their I/O addresses and interrupt usage are not very clearly defined. For this reason only the two fully standardized ports, COM1 and COM2, are considered.

Figure 4-1 is a simplified block diagram of the standard serial port. It consists of an INS8250 UART with RS-232 drivers and an interface to the PC I/O bus. A jumper set is provided to select the I/O address and interrupt (IRQ) line so that the port can be configured as either COM1 or COM2. The RS-232 lines are set up as a DTE device. The original IBM asynchronous adapter also provides a 20-mA current-loop interface, shown in Fig. 4-2. A jumper block on the card selects between RS-232 or current-loop operation. Third-party serial cards might not follow this original design. Some are direct clones; others are similar but omit the current-loop interface. Some offer options for configuration as COM3 or COM4 and might use a later version of the UART. Still others provide two ports on one card, or a serial and a parallel (printer) port. These variations are functionally equivalent in terms of programming and operation.

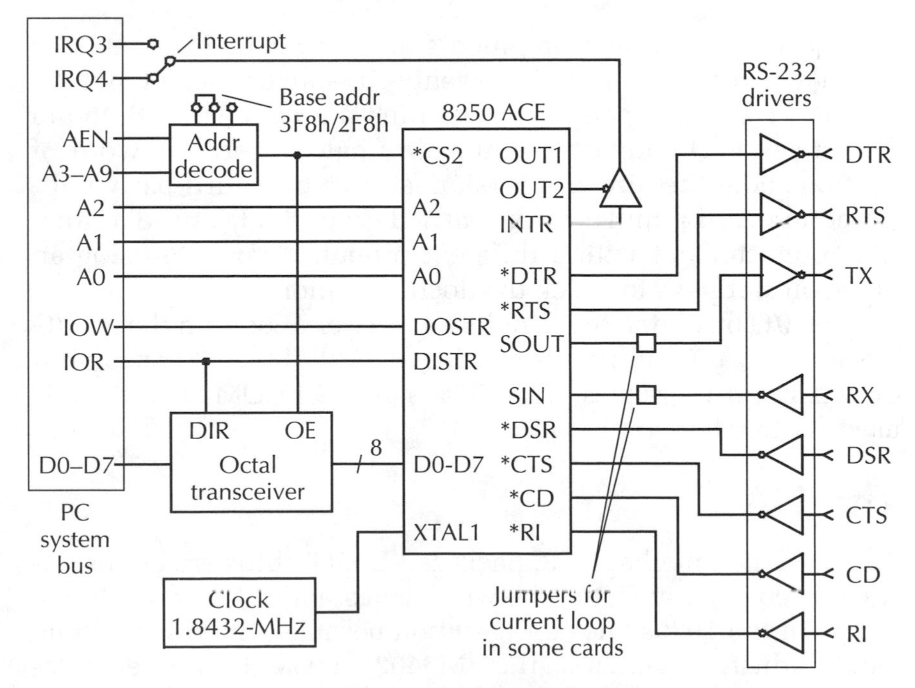

Fig. 4-1 *Block diagram of the standard PC serial port.*

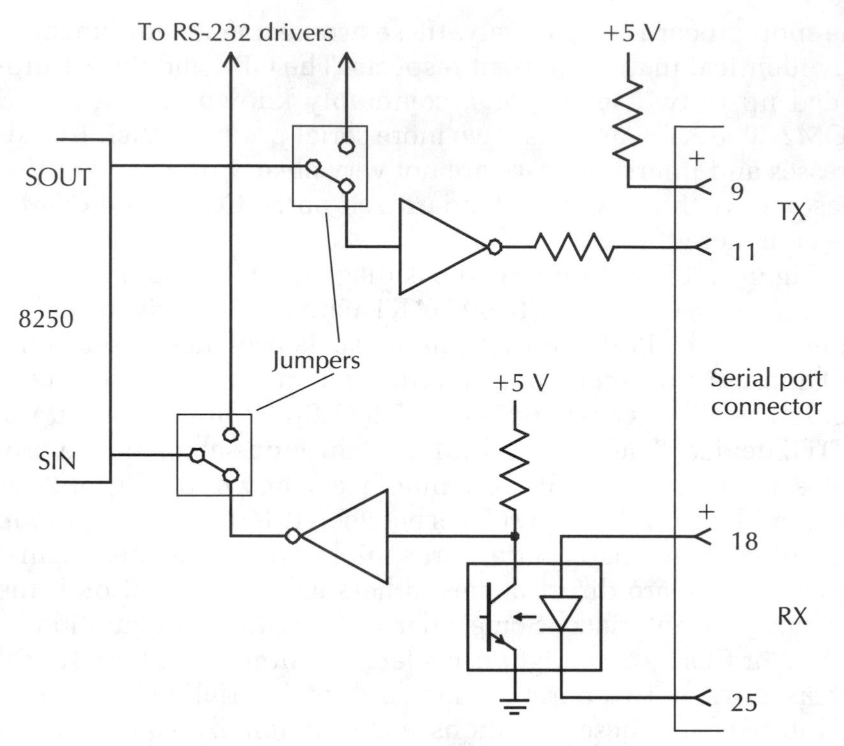

Fig. 4-2 *Current loop interface on the standard PC serial card.*

The RS-232 connector pinouts are shown in Fig. 4-3. The original IBM serial card uses a twenty-five-pin connector, but the AT (and many compatibles) use a nine-pin connector. Both are Canon series D connectors with pin (male) inserts. A word of caution about the nine-pin version: at least one third-party multifunction card popular in the early days of the PC used a nine-pin connector but with a different pinout. If you are using an older card, it pays to check the documentation.

The I/O bus interface circuits put the serial port on the PC I/O bus as a series of addresses starting at 3F8h if COM1, or 2F8h if COM2. Interrupt request line IRQ4 is used for COM1 and IRQ3 is used for COM2.

The UART

As mentioned in chapter 3, packaged UART chips are almost always used for translating between bytes and TTY serial frame codes. In the 1970s one configuration became something of a de facto industry standard (the IM6402, a CMOS implementation of this standard UART, is discussed in chapter 6). As micro-

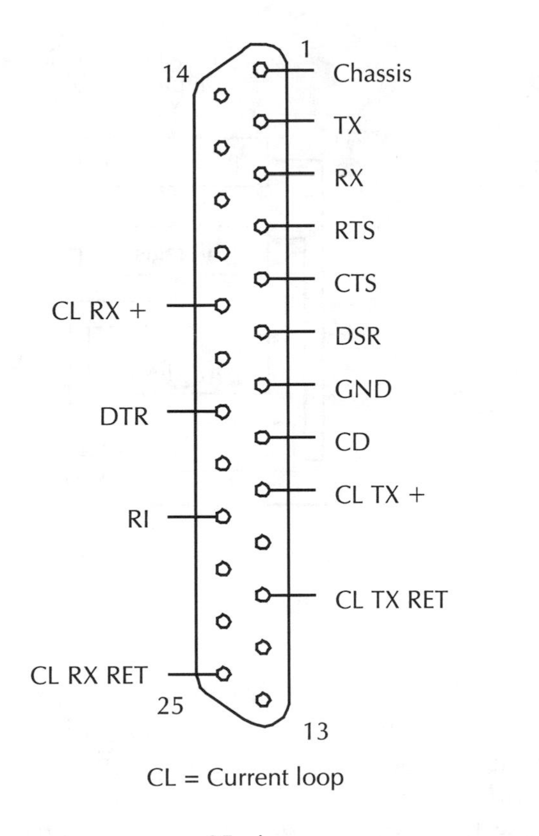

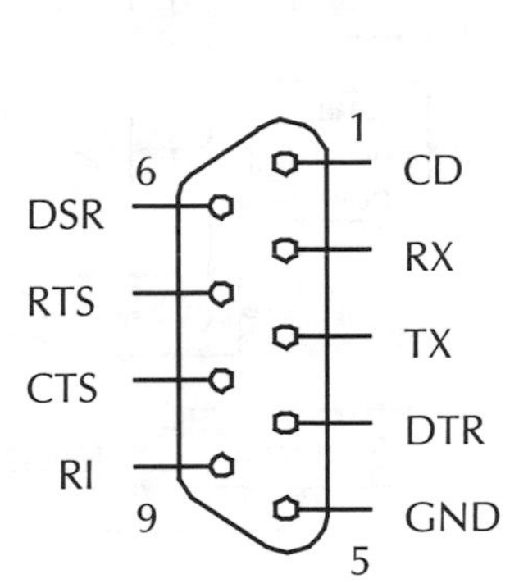

Fig. 4-3 *Serial port connector pinouts.*

processors became widespread, several UART designs especially intended for computer bus interfacing were developed. IBM chose one of them, the National Semiconductor INS8250, for the original PC. The AT uses a later enhanced version, the 16450, and the IBM PS/2 line uses a yet more recent variation, the 16500. Because these later chips amount to extensions of the 8250 they will not be considered separately.

The basic layout of a typical generic UART is shown in Fig. 4-4. It contains a parallel-to-serial converter for transmitting and a serial-to-parallel converter for receiving, basically the same circuitry as the simple shift register converters shown in Fig. 3-6. There are several other circuit blocks that provide various supporting functions. One of them is for phase-locking the receiver clock to the incoming start bit of each character, similar to the scheme shown in Fig. 3-12. Most modern UARTs actually define

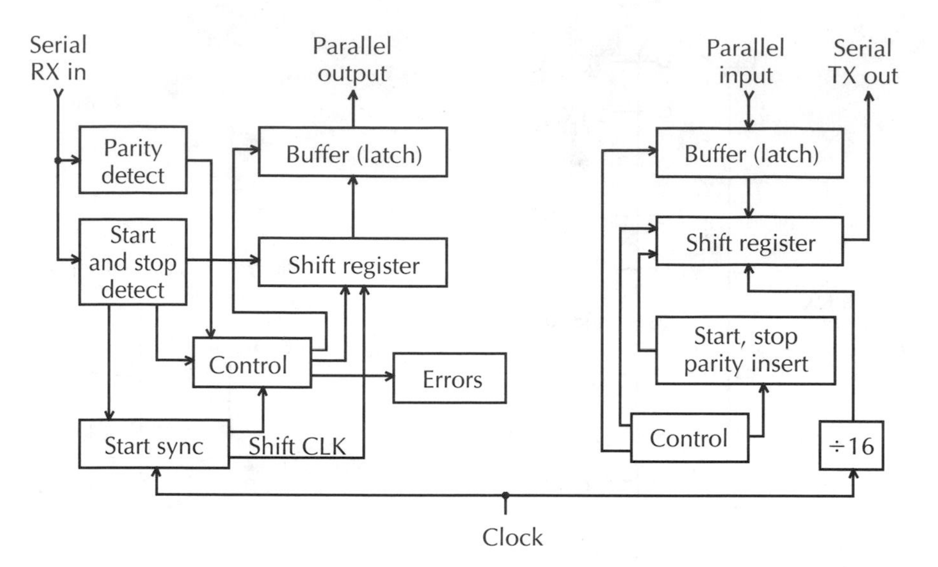

Fig. 4-4 *A hypothetical standard UART.*

the bit cell center to be at the half-cycle clock transition (that is, at 7½ clocks), giving better phase resolution. There is a set of buffers between the parallel data inputs and outputs and the shift registers. The buffers improve speed and efficiency by allowing the one byte to be put into the buffer while the shift register is working on another. The buffer size is commonly one byte, though some UARTs have more (for example, sixteen). Another supporting circuit block handles the insertion and recognition of parity bits and controls the generation of stop and start bits. Finally there are circuits that control the operation of the chip, enable or disable three-state outputs, handle error conditions, and other similar housekeeping tasks.

The INS8250 ACE

The 8250 extends the conventional UART design by adding a programmable baud rate generator, an interrupt controller, a set of modem control functions, and I/O ports expressly designed for bus operation. National calls it an Asynchronous Communications Element (ACE). Nearly all of its functions are controlled by writing command bytes into a set of registers in the device. All but one can be read from as well, which is very handy for saving the current state of the 8250 and restoring it later. Registers are selected by three address lines: A0 to A2. Reading from or writing to the 8250 registers is enabled by strobing *DISTR or *DOSTR respectively. Most of the registers use individual bits to control functions or report conditions. (A description of which bits do what in each register will be deferred until later.)

The 8250 registers at each register address (RA) are as follows:

- RA 0—data register (DR). When read from, this is the receiver buffer. When written to, it is the transmitter buffer ("holding register").
- RA 1—interrupt enable register (IER). The bits in this register enable or disable each of four internal "interrupt" levels within the 8250. These interrupts should not be confused with the PC interrupt system. The only thing they do is assert the INTR output of the 8250; a true CPU interrupt will occur only if the INTR signal is connected to the system bus. (The PC serial port can be set by software to connect or disconnect the INTR signal from the PC system bus.) The interrupt identification register (IIR) can be read to determine which 8250 interrupt level generated the interrupt. The four levels of internal "interrupts" in order of highest to lowest priority are
  (1) Receiver line status: interrupts if there is an overrun, framing, or parity error, or break. Reset by reading the LSR register.
  (2) Received data: a character has been received. Reset by reading the data register.
  (3) Transmitter holding register empty: the transmitter is ready for the next character to be sent. Reset either by reading the IIR register or by writing to the data register.
  (4) Modem status: interrupts if there is any change in the state of any of the modem control lines. Reset by reading the MSR register.
- RA 2—interrupt identification register (IIR). This register is read only. It indicates whether an internal interrupt is pending and, if so, which interrupt level caused it. If the INTR output of the 8250 is not being used to cause a CPU interrupt the 8250 "interrupts" are a useful status indicator, and reading this register is a simple way to check the current status of the device. (The 16500 UART also has a write-only register at RA 2, the FIFO control register [FCS], to control a sixteen-byte data buffer in the chip.)

- RA 3—line control register (LCR). The value in this register sets the serial format, the number of data and stop bits, and the type of parity if any.
- RA 4—modem control register (MCR). The bits in this register turn the four control outputs on and off. The 8250 provides DTR, RTS, and two general-purpose lines, OUT1 and OUT2. It also allows the device to be set to a "loopback" mode for testing. In this mode the transmitter is connected internally to the receiver and the modem control outputs are connected to the modem control inputs. (In the standard serial port, OUT1 is not connected to anything and OUT2 enables or disables a buffer that connects the 8250 INTR output to the IRQ line.)
- RA 5—line status register (LSR). The bits in this register report framing, overrun, and parity errors and the break condition.
- RA 6—modem status register (MSR). The bits in this register report the status of the modem control inputs DSR, CTS, CD, and RI. The upper four bits report the current state of these lines. The lower four bits are the so-called delta bits—they indicate whether the state of the corresponding line has changed since the MSR was last read. (The 8250 does not use the modem control lines in any way except to trigger a level 4 internal interrupt if so programmed.)

The 8250 contains an internal rate generator for clocking the transmitter and receiver. It is a sixteen-bit down counter that provides a divide-by-N of the chip clock input. Because the 8250 registers are one byte wide, two registers—the divider latch registers—are used to store the sixteen-bit divisor. Unlike the other registers, the divider latches cannot be accessed directly. Instead, bit 7 of the LCR—the divider latch access bit (DLAB)—must first be set to 1. The divider lsb latch can then be written to or read from at address R0, and the msb latch likewise at address R1. Clearing the DLAB bit to 0 restores the 8250 to its normal configuration.

There is, by the way, a bug in early 8250 devices such that the divider latches will load only when the chip is in loopback mode. It is a good idea to set the 8250 to loopback before setting the baud rate just in case you are dealing with one of these early chips.

Programming the PC serial port

Programming the standard PC serial port is mostly a matter of programming the 8250 UART. The register addresses of the 8250 are mapped directly to PC I/O addresses starting at a base address of 3F8h for COM1 and 2F8h for COM2. The actual addresses are shown in Table 4-1, but in programs it is usually more convenient to compute the addresses as base + RA. The base addresses for COM1 and COM2 have remained unchanged throughout the history of the IBM PC.

Table 4-1 Serial port register addresses.

8250 register	COM1	COM2	Name
0	3F8h	2F8h	Tx, Rx
1	3F9h	2F9h	Interrupt enable (IER)
2	3FAh	2FAh	Interrupt identification (IIR)
3	3FBh	2FBh	Line control (LCR)
4	3FCh	2FCh	Modem control (MCR)
5	3FDh	2FDh	Line status (LSR)
6	3FEh	2FEh	Modem status (MSR)
With LCR bit 7 (DLAB) = 1			
0	3F8h	2F8h	Divisor lsb
1	3F9h	2F9h	Divisor msb

Configuring the serial port

The hardware of each serial port must be set up so that the port uses the proper I/O addresses and interrupt request lines. Standard serial cards usually include jumpers or switches to configure the port as COM1 at I/O address 3F8h or COM2 at 2F8h. Usually the port I/O address and the interrupt request line are selected by separate switches or jumpers; both must be set correctly. IRQ4 is ordinarily used for COM1 and IRQ3 for COM2. In many recent computers, at least one port is built in to the motherboard and the configuration jumpers or switches are located there. Plug-in modem cards must likewise be set to act like a COM1 or COM2 device.

Two additional serial ports—COM3 and COM4—were introduced with the IBM AT. Unfortunately the situation with regard to these ports is somewhat murky. Cards with additional ports seem to use a variety of addresses. Address 3E8h seems fairly common for COM3 and 2E8h for COM4, but there are exceptions. A more substantial problem is the use of interrupt request lines. Two devices cannot share the same IRQ line. If IRQ4 and IRQ3 are already being used by COM1 and COM2 ports, some other IRQ line must be used for a COM3 or COM4 interrupt. IRQ2 is the only line that can safely be commandeered for this purpose in an XT (unless you are willing to give up the printer interrupt on IRQ7), and even it might already be used by some other accessory device such as a LAN card. In principle, COM3 and COM4 are defined for AT (sixteen-bit bus) machines, so the reasonable solution is to use one of the additional IRQ lines in the sixteen-bit bus if the serial adapter supports it (eight-bit cards, of course, cannot). This has repercussions for software. As noted in chapter 2, interrupt service routines for these additional interrupts must control both the master and the slave interrupt controller.

Programming the PC serial port

The serial port is programmed by writing the appropriate values to one or another of the UART registers. The I/O port address for the registers is conveniently expressed as (base + n), where n is the UART register address (RA) and base is the serial card base address (3F8h or 2F8h). The actual addresses are listed in Table 4-1.

Communication parameters The number of data and stop bits and the parity bit are set through the line control register, port (base + 3). The bits are as follows:

- bits 1, 0: the number of data bits, where 00 = 5 bits, 01 = 6 bits, 10 = 7 bits, and 11 = 8 bits.
- bit 2: the number of stop bits, where 0 = 1 bit and 1 = 2 bits (or 1½ bits if 5 data bits are selected).
- bits 5, 4, 3: the type of parity, where 000 = no parity, 001 = odd, 011 = even, 101 = mark, and 111 = space parity.
- bit 6: sets break where 0 = off, 1 = on.
- bit 7: the baud generator divisor latch enable (DLAB) bit, where 0 = normal operation and 1 = load divisor bytes.

Baud rate As with most UARTs, the transmitter and receiver of the 8250 are clocked at sixteen times the baud rate. To get a particular baud rate R, set the 8250 to divide the chip clock CLK by a divisor D such that $CLK/D = 16R$, or $D = CLK/16R$. Because CLK in the PC is fixed at 1.8432 MHz, this can be simplified to $D = 115200/R$.

Because the sixteen-bit divisor D can range from 1 to 65,535 (FFFFh), baud rates can, in principle, run from 1.76 to 115,200. Recent versions of the 8250 and its successors often will run at 115 kilobaud, but this cannot be guaranteed. The 8250 data sheet specifies a maximum rate of 56 kilobaud, which implies that the minimum safe D is 2, giving 57,600 baud. Remember, too, that the divisor is an integer. At the higher rates where D is small, the selection of rate becomes rather coarse. The next lower rate below 57,600, for instance, is 38,400 ($D = 3$). Fortunately, the most commonly used rates are available.

Because the rate divider latch has to be loaded indirectly, setting the baud rate involves several steps:

1. Calculate the required integer divisor D = 115,200 div R (div denotes integer division). If floating-point division is used, the quotient must be rounded or truncated to an integer.
2. Set bit 4 of the MCR to 1 to turn the loopback on (for the sake of early 8250s). This can be done by writing 10h to I/O port (base + 4).
3. Set bit 7 of the LCR to 1 to turn DLAB on in order to have access to the divider latch. This can be done by writing 80h to port (base + 3).
4 Write the least significant byte of D to port base. It will go into the divider latch.
5. Write the most significant byte of D to port (base + 1), the IER. It will go into the divider latch.
6. Clear bit 7 of the LCR to 0 to turn DLAB off and restore normal register operation. This is a good time to write the communications parameter byte (data bits, stop bits, parity) to the LCR. Write it to port (base + 3).
7. Clear bit 4 of the MCR to 0 to turn the loopback off. This can be done by writing 0 to port (base + 4). The 8250 is now ready to go. The modem control lines will be in the default off state.

Control lines The 8250 has four output lines (DTR, RTS, OUT1, and OUT2) that are controlled by bits in the modem control register, port (base + 4). These outputs are for convenience and are not used internally by the UART. The MCR bits control the lines as follows (clearing a bit to 0 turns that line off, and setting it to 1 turns it on):

- bit 0: DTR
- bit 1: RTS
- bit 2: OUT1
- bit 3: OUT2
- bit 4: controls the loopback mode. Clearing bit 4 to 0 selects normal operation, and setting bit 4 to 1 forces loopback.

The PC serial port connects the 8250's DTR and RTS outputs through RS-232 drivers to the serial port output connector for use as regular modem control lines. OUT2 controls a three-state buffer that connects the 8250 INTR pin to an IRQ line on the PC bus. If OUT2 is set to 0, the IRQx line will not be activated and the PC serial interrupt will be disabled (the 8250 internal "interrupts" are not affected). To run the serial port with interrupts, OUT2 must be set on (bit 3 = 1).

In standard PC serial cards OUT1 is not connected to anything, but plug-in modem cards often use it to control modem reset.

Status registers The state of most of the functions of the serial port can be determined from the two status registers of the 8250: the line status register (LSR) and the modem status register (MSR). Each bit in the status registers indicates the current state of a particular function. To determine the state of the bit, read the byte from the register and AND it with a mask with the desired bit set to 1 and all other bits set to 0. If the result is not zero, the bit was set.

1. Line status, port (base + 5). The bits of the LSR reflect the state of the transmitter and receiver and also report any receiving errors that occurred in assembling the byte.

   - bit 0: set = a byte has been received and is waiting
   - bit 1: set = overrun
   - bit 2: set = parity error
   - bit 3: set = framing error
   - bit 4: set = break
   - bit 5: set = transmitter ready for the next byte to send
   - bit 6: set = transmitter shift register empty

2. Modem status, port (base + 6). The lower four bits are set if the state of any of the control lines has changed since the

MSR was last read. The upper four bits reflect the current state of the MSR at the moment it is read. In the following list, a high level (that is, the line is ON) refers to the state of the external RS-232 lines, not the actual 8250 TTL-level input pins.

- bit 0: set = delta-CTS
- bit 1: set = delta-DSR
- bit 2: set = delta-RI
- bit 3: set = delta-CD
- bit 4: set = CTS input is high
- bit 5: set = DSR input is high
- bit 6: set = RI input is high
- bit 7: set = CD input is high

Internal interrupts The internal interrupts are enabled and disabled by bits in the interrupt enable register, port (base + 1). Remember, these are not part of the PC interrupt system; they affect only the INTR output pin of the 8250. A PC interrupt will be generated only if OUT2 is turned on to connect the 8250 INTR pin to the PC bus.

Each of the four internal interrupts is enabled by setting the corresponding bit to 1 or disabled by clearing it to 0:

- bit 0: received data
- bit 1: THRE
- bit 2: line status
- bit 3: modem status

The interrupt identification register, port (base + 2), can be read to determine the status of internal interrupts. If bit 0 = 1, there are no internal interrupts pending. If bit 0 = 0, then bits 1 and 2 indicate which level generated the interrupt:

- 00: change in modem status
- 01: transmitter buffer empty (THRE)
- 10: data received
- 11: line status (error or break)

Is anybody there? (Checking for a valid port)

If you are running software on a particular computer that you are familiar with, you probably know which serial ports have been installed. In general, though, it's not a very good idea to blithely do I/O to a serial port on the mere assumption that it exists.

There are two ways to check for the presence of a valid serial port. Both are illustrated in the short program in Fig. 4-5. The simplest and fastest and in many respects the best way is to examine the list of port addresses in the BIOS data area. They are in an array of four sixteen-bit words starting at address 40:0 hex. Each entry corresponds to one of the ports from COM1 to COM4. Nonzero values are the base I/O address of the respective ports; a zero indicates that no hardware is installed for that port. The function **CommBase()** in Fig. 4-5 shows how to read the array. The computation of the index to the array is shown in expanded form for clarity; a more concise expression would be Table[(*N*-1) and 3]. (A utility for listing all serial and parallel ports is given in chapter 7.)

The BIOS table method might fail in certain special cases. If the computer is being used as a network file or printer server, for example, the network shell might set some or all of these port base addresses, and the similar ones for parallel ports, to zero even though the hardware exists. The idea is that these ports are, or could be, used for network printers and therefore should be marked as nonexistent so that applications will not try to use them.

Fig. 4-5 *Routines to check for the presence of a serial port.*

```
{ Example of how to get serial port address from BIOS table. }

function CommBase( N : byte ) : word;                    { 1 = COM1, etc. }
var
   Table : array [0..3] of word absolute $40:0;
begin
   N := (N-1) and 3;                                   { adjust N and clamp }
   CommBase := Table[N];                                       { get value }
end;


{ Routine to verify presence of serial port by hardware test. }

function TestCommHdwe( Base : word ) : boolean;
const
   MCR = 4;
   MSR = 6;
var
   B1,B2,S : byte;
begin
   if Base <> 0 then
      begin
         S := Port[Base+MCR];                            { save MCR status }
         Port[Base+MCR] := $11;                        { set loopback + DTR }
         B1 := Port[Base+MSR] and $F0;                     { DSR bit = $20 }
         Port[Base+MCR] := $12;                        { set loopback + RTS }
         B2 := Port[Base+MSR] and $F0;                     { CTS bit = $10 }
         Port[Base+MCR] := S;                             { restore status }
         TestCommHdwe := (B1 = $20) and (B2 = $10);      { exists if equal }
      end else TestCommHdwe := false;
end;
```

The second method is to test the hardware directly by attempting to set a device to a certain state and seeing if it responds. This is how the BIOS determines valid port addresses for the port data table. The function **TestCommHdwe()** in Fig.4-5 tests for hardware by attempting to set the UART to loopback, writing to the modem control port, and then reading the modem status port. In loopback mode the control bits directly change the status bits, so if the status bits assume the expected values, the port presumably exists. The routine uses DTR to set DSR, and then RTS to set CTS. This double-barreled approach is probably overkill. On the other hand, it is exceedingly unlikely that any device other than a serial port would respond in the expected way.

TestCommHdwe() also illustrates a good way to read and write to the 8250. The 8250 register addresses are declared as constants. The I/O addresses are then calculated from the variable *Base* plus the appropriate register address. The same routine can be used for any of the serial ports simply by assigning the appropriate value to ***base***. Turbo Pascal does I/O to ports by treating I/O space as an array; assignments to or from **Port[ ]** compile to an in or out. In some other languages port I/O is expressed as a function; for example, **inportb()** and **outportb()** in Turbo C.

Using the BIOS table is ordinarily the method of choice. The hardware method is shown largely for illustration. It could, however, be expanded into a useful general diagnostic utility for checking 8250 chips. Another useful utility would be one in which the modem control lines are turned on and off in response to keyboard commands. For some unfathomable reason the connectors on the back of computers seldom bear labels. By manipulating, say, the DTR line, the serial port or ports could be identified with the help of a voltmeter.

Polled serial port I/O

The difference between polled and interrupt-driven I/O was introduced in chapter 2. In polled operation you write bytes to, and fetch bytes from, an I/O port. It is much like general read and write operations. Most of the code examples presented so far in this chapter are of this kind. Little needs to be said about input and output functions required in polled operation. Serial port output functions write a byte to the serial port data register (an OUT to the port base address). Input is simply an IN from the data port.

The key to successful polled operation of any I/O device is

knowing its current status. It won't do to write to it or read from it blindly; you need to be able to check if the device is ready and to find out if it has any new information waiting. In the case of the serial port the line status register has the information you need. The bits in the status byte indicate (1) if there is a received byte waiting to be read, (2) if the transmitter is ready to accept the next byte to be sent, and (3) whether there have been any receive errors or a break. To determine the serial port status read from address (base + 5) and check the appropriate bits. For example:

1. Is the transmitter ready for another byte? Yes, if bit 5 of the LSR is set: that is, if port (base + LSR) and $20 <> 0. (The notation in Pascal for not equal is <>; in C it is !=.)
2. Has a byte been received? Yes, if bit 0 of the LSR is set: that is, if port (base + LSR) and 1 <> 0.
3. Have any errors occurred? Yes, if one or more of these bits are set:

   - bit 1: overrun (LSR and 2 <> 0)
   - bit 2: parity (LSR and 4 <> 0)
   - bit 3: framing (LSR and 8 <> 0)
   - bit 4: break (LSR and 10h <> 0)

The status of the modem control lines can be checked by examining the bits of the modem status register (MSR) byte obtained by reading from address (base + 6). The lines can be turned on and off by setting or clearing the appropriate bits in the modem control register (MCR) at address (base + 4). Remember that bit 3 of the MCR must be zero for polled operation so that the port is disconnected from the PC's interrupt system.

Example of a polling routine

The skeleton of a routine for handling polled serial I/O is shown in pseudocode in Fig. 4-6. **Receive-data-waiting** and **transmitter-ready** are functions that check the status of the serial port and return a boolean value. **Serial-error** reads the LSR and checks for errors. The items shown in brackets concern flow control. If software flow control is used, it is processed as part of the received data. If hardware flow control using a modem status line is in effect, the MSR should be read and checked (shown as a test for BUSY in Fig. 4-6). These functions can be implemented in the manner suggested in the previous paragraph. Note that the transmit operations are placed last in the loop. This ensures that any

Fig. 4-6 *Example of a "communication engine."*

```
procedure Poll:
while not Quit do
   begin
      if receive-data-waiting then
         begin
            read serial port data register
            process received byte
            [handle flow control codes]
         end
      if serial-error then
         handle error
      [if BUSY then
         set transmit pause]
      if key-pressed then
         begin
            read keyboard
            process keyboard character
         end
      if byte-to-send then
         if transmitter-ready then
            write byte to be sent to serial port data register
   end while
end Poll
```

flow control codes or keyboard entries to suspend transmission are already handled.

A status function is also shown for the PC keyboard. It returns TRUE if any keyboard characters are waiting to be read. Most compiler libraries offer such a function; Turbo Pascal provides **KeyPressed** and Turbo C has **kbhit.** Otherwise the keyboard status can be obtained by calling either MS-DOS function 0Bh (via Int 21h) or the BIOS keyboard service routine, interrupt 16h.

In many applications, other input and output devices would be incorporated in the polling loop also. For example, a program could accept input from a pointing device such as a mouse. A polling loop would get the current status from the mouse driver (through interrupt 33h for Microsoft and compatible mice) and act accordingly. Because the Microsoft mouse driver allows user routines to be chained to the mouse driver interrupt system, another method would be to write a mouse handling routine that would among other things control a flag variable in the polling loop. Yet another approach would be to use a chained mouse interrupt handler to put mouse events in a buffer in a manner similar to the serial port input buffer to be described later in this chapter.

Routines like the one in Fig. 4-6 have been used for many years and have even acquired names like "communication engine." It is interesting, and perhaps a little sobering, to realize that such a control structure implements event-driven program control. The set of status functions constitutes a small and simple but nevertheless genuine event queue. (Indeed, this sort of control structure has long been referred to as a "dispatcher" or "scheduler.") Event-driven programming has become something of a buzzword of late, but it is really not so new. The distinction between event-driven and procedural control is not quite as watertight as one might think.

Another way to implement a polling loop for an 8250 is to use the chip's internal interrupt system as a status indicator. The 8250 is disconnected from the PC IRQ line by turning OUT2 off (clear bit 3 of the MCR to 0), and the internal interrupts in the 8250 are enabled by setting the appropriate bits in the IER. The serial status functions in the polling loop are replaced by ones that examine the IIR instead. If bit 0 of the IIR byte is set to 1, no interrupts are pending; if zero, the other bits are examined to see what needs to be done. This is especially handy for implementing very tight polling loops in assembly language. An IN from (base + IIR) puts the IIR byte in AL which is then rotated right through carry (RCR), putting bit 0 of the IIR in the carry flag. If carry is set, jump to the top of the loop. Otherwise the remaining bits in AL can be used as an index to offsets of subroutines to handle receive (if AL = 2), transmit (AL = 1), modem status (AL = 0), or errors (AL = 3) as the case may be.

The fatal flaw in polling schemes is not an inability to check status fast enough—a tight loop can keep up with an 8250 to fairly high baud rates—but the time required by other routines to handle bytes that are received or to prepare the bytes that are to be transmitted. What is needed is a means to handle the received bytes immediately whenever they come in. To do that effectively requires the use of interrupts.

Interrupt routines

Interrupts were introduced in chapter 2. You saw there that an interrupt routine is very much like an ordinary subroutine. The major difference is that ordinary subroutines are invoked with the CALL instruction, while interrupt routines are invoked either by the INT instruction (software interrupts) or electrically by raising the INTR input pin on the CPU to high level (hardware interrupts). Other differences are that the CPU saves the flags as well as the return address before executing an interrupt routine, the routine must terminate with an IRET instruction rather than an ordinary return, and interrupt routines are always far calls. Let's turn now to the software aspects of interrupts and see how an ISR is actually written.

An ordinary subroutine in a high-level language must do its stuff without affecting other parts of the program unless that is specifically intended. That is, the subroutine must preserve the information needed to ensure that program execution can continue properly after the subroutine is finished. In high-level languages the compiler takes care of these details automatically. In an interrupt routine you must do so yourself, explicitly. The first and greatest rule in writing interrupt routines is that *the necessary system state information must always be preserved.*

Saving the CPU state

What is the state information? The answer depends on whether an ISR is intended to be invoked by software or hardware. A software interrupt will (presumably) be called explicitly and under known and appropriate conditions. The CPU registers are commonly used to pass information between interrupt and caller, and which registers will contain what can be specified in advance. The calling routine can take into account any program variables or CPU registers that the ISR might modify. A hardware interrupt routine, in contrast, may be invoked at anytime. Consequently a hardware ISR must make no assumptions about the CPU registers, and when it is finished it must ensure that they are exactly as they were when the ISR began. In short, a hardware interrupt routine must be self-contained and completely self-sufficient. The only thing an ISR "knows" when it is invoked is its own location in memory (at CS:IP).

In practical terms this means that the first thing an interrupt routine must do is to save all of the CPU registers it will modify, and then restore these saved values as its last action when it is done. Furthermore, it must contain within itself the means to access any external variables it uses. This is simple if such variables are at known locations (like the video buffer or the BIOS data area), but if they are in an application program, some way has to be provided for the ISR to find out where the program's data area is.

There are several ways in which the CPU registers can be saved. Because the interrupt routine knows its own location in memory, an area can be set aside within the routine in which to save the CPU registers. Another approach is to save registers on the stack. Because some readers might not be familiar with how the stack works, a brief explanation seems in order.

Very often in the course of computations there is a need for temporary storage of intermediate results and the like and to preserve certain system variables such as CPU registers—in short, a scratch pad. One method is to set up an array to hold bytes temporarily with an index variable to indicate the next available free space in the array. Each time a byte is added, the index is incremented. Logically this is analogous to stacking items one on top of the other, and the array index points to the current "top" of the stack. Adding an item and incrementing the index is usually called *pushing*, and retrieving the topmost item and decrementing the index is called *popping*. Access to items within the stack is gained by simple arithmetic: the third item from the top is at location (index – 3). This is such a good way of doing things that in nearly all modern computers the CPU includes special stack-management capabilities that allow the stack area to be addressed as a separate entity (a stack segment) and that manipulate the index (stack pointer) automatically in response to PUSH and POP instructions. High-level languages make extensive use of the stack to pass data to and from subroutines and to hold the local variables of subroutines. In the 80*x*86 and some other CPUs, the stack is laid out in memory such that added items are at lower memory addresses; the stack "grows downward" (it is a stalactite rather than a stalagmite). The address of the *N*th item from the current top (bottom?) is thus (stack pointer + *N*). (The 80*x*86 also provides an auxiliary stack pointer register (BP) that can be set to SP at the beginning of a subroutine to simplify the creation of, and access to, local variables.)

An application program ordinarily declares an area of memory to be used for a stack. The exception is COM programs in which code, data, and stack overlap. When the program is loaded, DOS sets the CPU stack registers SS and SP to this area. Although it is seldom done, a program could also set up additional stack areas of its own. (One program that does so is MS-DOS itself which maintains at least three internal stacks.) Now when an ISR receives control, the CPU stack registers are still set for whatever stack area was in effect when the interrupt occurred. This is not necessarily the program's stack; the interrupt could have occurred during a DOS operation, for example. An interrupt routine thus has no way of knowing how much stack space is available to it. Accordingly, an ISR that uses the stack for temporary storage should either use as little space as possible or set up its own stack. Most small interrupt routines, such as the ones in this book, choose the former alternative.

The situation changed beginning with version 3.2 of MS-DOS. These versions of DOS intercept hardware interrupts and switch the CPU to a spare stack before passing control to the interrupt routine, and switch it back after the interrupt routine is finished. By default there are nine spare stacks of 128 bytes each; this can be changed by means of a STACKS= command in the CONFIG.SYS file.

Accessing data from an ISR

In most cases an interrupt routine will need to access data that is located somewhere outside of the routine. If the data is at a known and fixed address, such as video RAM, this is straightforward. Most of the time, however, the data will be at an unknown location, typically somewhere in the application program that is running. In the module of serial routines described later in this chapter, for example, the base address of the serial port needs to be fetched and the received character must be put in a buffer in the module's data area. This can be done in several ways, depending on the way that the program and the ISR are laid out in memory.

In standard COM files, which use the so-called tiny memory model, code and data overlap. Data items can be accessed directly with the CS: segment override prefix. If the DS register is to be used, it should be saved and then loaded with the current value of CS because the contents of the DS register might be anything when the interrupt occurs. (Versions 3 and earlier of Turbo Pascal generate COM files but maintain a "data segment"—actually a secondary heap—for global variables accessed through DS. Typed constants are stored in the code segment. The manuals explain how to access global variables from within an ISR.)

EXE files allow separate code, data, and stack segments. The file contains a header that identifies segment references in the program, and MS-DOS plugs in the actual address information (segment fix-ups) as it loads the program into memory. By requesting a data segment fix-up as part of its code, the interrupt routine can get the proper address to put into DS to address the program's data segment. This method is used in the examples in this book. Note that Turbo Pascal versions 4.0 and above use the "medium" memory model with a data segment, a stack segment, and two or more code segments. All major C compilers support all of the standard memory models.

Accessing other routines

Sometimes it would be useful to call other subroutines from within an interrupt routine. If those subroutines are part of the ISR itself, there is no problem. But what about subroutines elsewhere, such as in the main program? Because an interrupt can occur at any time, the question arises: what if that subroutine is already "in use" by the program when the interrupt occurs? To answer you must first consider the important matter of *reentrancy*.

It is revealing to classify subroutines according to the degree to which they can be called more than once, or "reused." Consider first the case in which an ordinary subroutine in a program uses global variables but does not initialize them. Each time the subroutine is called the global variable will have a different value, and the subroutine will change it further. Such a subroutine is said to be nonreusable. Sometimes this is exactly what is wanted. Variables that accumulate some sort of running value are a prime example. The CRC update routines in chapter 3 are written this way so that each call updates the CRC accumulator. Recall that an explicit initialization is required each time a new CRC is started.

Now consider a subroutine that uses global variables but initializes them each time it starts work. Each time it is called its calculations start off from the same place, so to speak. If it is fed the same parameters with each call, it will return the same result. Such a subroutine is *serially reusable*.

Yet another possibility is a subroutine arranged so that a group of variables is set up and initialized each time it is called (and discarded after it is finished to save memory). These variables are "local"; they can be accessed only by the subroutine itself. If such a subroutine is called again while it is in the midst of a calculation, the second invocation will also create a new group of local variables for the second instance of the subroutine. Because the variables of one instance of the subroutine do not affect the variables of another, the subroutines remain independent. Such subroutines are *reentrant*. They can be interrupted at any point and still continue on to a correct result when they resume. A special case of reentrancy is when a subroutine can call itself; such subroutines are *recursive*. The subroutines of most modern procedural languages, including Pascal and C, are fully reentrant and support recursion (provided that they use local variables). The stack is used to store the parameters and local variables separately for each invocation of the subroutine.

It follows that it is safe for an ISR to call some other subroutine only if that other routine is reentrant. In most modern high-level procedural languages it is safe to call subroutines you have written, and also many (usually not all) of the compiler library subroutines. However, the ROM BIOS and DOS are not reentrant; they are serially reusable. Because an interrupt can occur at any time, a BIOS or DOS routine might already be in progress when the interrupt occurs. It is therefore not permissible to call BIOS or DOS routines—or any language routines that in turn use them, such as for file or keyboard operations—from within an ISR. (MS-DOS function 34h returns a far pointer in ES:BX to the InDos flag. An ISR can in principle use this pointer to examine the InDos byte, which is 0 if a DOS function is not currently active. Microsoft does not guarantee that this will be valid for all versions of DOS nor under Windows. It is best not to call DOS from an ISR under any circumstances.)

Summary A lot of ground has been covered in the last paragraphs, so a brief summary seems in order. An interrupt service routine (ISR), you have seen, is a far subroutine that must have a certain structure and do certain things. Specifically,

1. The first thing an ISR must do is save the system state—basically, any CPU registers that the ISR will change in any way—either on the stack or within itself.
2. If the ISR will access any external data items, which is usually the case, the means to address such data must be set up.
3. The ISR can now begin its own work. The ISR may call other subroutines outside of itself, provided that they are reentrant.
4. In the case of a hardware interrupt, the ISR must reset the interrupt controller (see the next section).
5. The ISR must restore all saved registers.
6. The ISR must conclude with an IRET instruction.

Some high-level languages such as Turbo Pascal and Turbo C provide a special subroutine type—**interrupt**—that generates code to take care of items 1, 2, 5, and 6 for you automatically. This is quite convenient. Some of the code examples in this book use this handy feature.

What has been said so far covers the ground needed for software ISRs —those invoked by the 80x86 INT instruction—but there are some additional considerations that apply to ISRs invoked by hardware.

Hardware interrupts and the interrupt controller

Interrupt routines that have been invoked by hardware (a high signal on the INTR input pin of the CPU) start off with the interrupt flag cleared, which suppresses further maskable hardware interrupts. If the ISR is not short and fast, it is usually a good idea to issue an STI instruction (FBh) to enable other interrupts. They can be turned off again with a CLI instruction (FAh) if needed.

Hardware interrupts also involve the interrupt controller (PIC). To begin with, it must be set up initially so that it will respond to the particular interrupt request line to be used. Moreover, it must be reset every time it is triggered by an interrupt. The structure and operation of the 8259A PIC used in PC systems was covered in chapter 2; here the programming side of the picture is discussed. I will deal only with those aspects of programming the 8259A that directly affect interrupt routines. The 8259A has many modes of operation, and the configuration set up by the BIOS during initialization should not be changed unless you know exactly what you are doing and have taken into account any side effects on the BIOS or other system resources.

The end-of-interrupt command When the PIC initiates a hardware interrupt it blocks further interrupt requests on that IRQ line and all lines of lower priority until it is reset. Every ISR must therefore issue an end-of-interrupt (EOI) command to the PIC before it concludes. (This is item 4 in the structural outline of an ISR given previously.) Ordinarily the "nonspecific" EOI command is used, 20 hex (32 decimal), written to the PIC at its base I/O address. For IRQ0 to IRQ7 the base address of the PIC is 20h in all cases (OPC, XT, and AT machines). In the case of IRQ8 to IRQ15 in AT machines, the IRQ line activates the slave PIC which in turn activates IRQ2 on the master PIC, and both PICs must be reset with an EOI command. The base address of the slave PIC is A0h. These items are summarized in Table 4-2.

The interrupt mask In order for an IRQ line on the PC bus to generate an interrupt, the corresponding input of the PIC must be set to the "enabled" state. The PIC uses the eight bits in a command byte (OCW1) to enable or disable each of its eight interrupt request inputs. Bit 0, the lsb, controls IRQ0; bit 1 controls IRQ1; and so forth. Note that each IRQ input is enabled by clearing the corresponding bit to zero, and disabled by setting it to one. A mask byte is put into the PIC by writing it to I/O port (base + 1).

Table 4-2 Interrupt controller commands.

Machine type	Action
PIC base I/O addresses	
OPC, XT	20h
AT, 386	Master, 20h Slave, A0h
Nonspecific EOI reset (EOI = 20h).	
OPC, XT	Write EOI to port 20h
AT, 386	IRQ 0–7 Write EOI to master, port 20h IRQ 8–15 Write EOI to slave, port A0h Write EOI to master, port 20h
IRQ mask (bit = 0 to enable, 1 to disable).	
OPC, XT	Write mask to port 21h
AT, 386	IRQ 0–7 Write mask to port 21h IRQ 8–15 Write mask to port A1h

As shown in Table 4-2, this is port 21h for the OPC, XT, and AT master PIC, and port A1h for the AT slave PIC. The current mask can be examined by a read from the same port.

The interrupt mask should be set carefully because it controls all of the IRQ inputs of the PIC. For instance, accidentally disabling IRQ1, the keyboard interrupt, causes the keyboard to go dead. The best way to avoid trouble is to fetch the current mask, modify the bit corresponding to the interrupt you are setting, and then replace the mask. It is also a good idea to define a control byte for the IRQ you want to manipulate. The bit in the control byte corresponding to the IRQ number you want is set to 1 and all other bits are 0. You can then use the control byte to enable or disable that IRQ number as follows:

To enable:

1. Get the current mask: `Mask := Port[PIC_Base+1];`.
2. Clear the relevant bit: `Mask := Mask and not ControlByte;`.
3. Load the modified mask: `Port[PIC_Base+1] := Mask;`.

To disable:

1. Get the current mask: Mask := Port[PIC_Base+1];.
2. Set the relevant bit: Mask := Mask or ControlByte;.
3. Load the modified mask: Port[PIC_Base+1] := Mask;.

Installing and removing interrupt routines

In order to use an interrupt routine it must be installed; that is, its entry point address must be put into the proper slot in the interrupt vector table. This should always be done through DOS, not by writing directly to the table, especially if a network shell, DOS extender, or Microsoft Windows is running. MS-DOS provides function 25h to set an interrupt vector and a similar function 35h to get a vector.

To put an ISR entry point in the table:

Call Int 21h with
    AH = 25h
    AL = vector number (0–255)
    DS = ISR address segment
    DX = ISR address offset

To get an address from the table:

Call Int 21h with
    AH = 35h
    AL = vector number (0–255)
Returns
    ES = address segment
    BX = address offset

High-level language compilers for the PC often provide utility routines that implement these calls. In Turbo Pascal they are **SetIntVec** and **GetIntVec**, and in Turbo C they are **setvect** and **getvect**. Both take as parameters the interrupt number, a byte 0–255, and a (far) pointer that holds the address.

It is good programming practice to leave things the way you found them whenever you modify any aspects of a computer system. (You do that, don't you?) In the case of interrupt routines it is mandatory. If a program exits but leaves an interrupt vector that it installed in place, the address that the vector points to will no longer be valid. When another program is loaded, that location in memory will doubtless contain something quite different. If the interrupt is triggered for any reason, including noise, the system will try to execute whatever is at that address. By far the most likely result is a system crash.

A program that installs an interrupt routine should first get the current vector and save it, and only then install the address of the ISR in its place. Before the program exits back to DOS, it should replace the saved original interrupt vector.

One way to do this is to place code to restore the saved interrupt vector among the cleanup routines the program executes just before it exits back to DOS. But suppose the program terminates prematurely due to a fatal run-time error, such as an attempted divide-by-zero. The exit cleanup, including the restoration of the original interrupt vector, will not be executed. Fortunately, there is a way around this problem. Nowadays the run-time kernel of many high-level languages installs its own exception handlers and provides a means of linking exit procedures into an orderly shutdown process. Turbo Pascal lets a program insert exit procedures in a linked list using the pointer *ExitProc*; the Serial Toolkit module later in this chapter illustrates the method. In Turbo C the **exit()** function will call exit routines that have been registered through the **atexit()** function. Remember, ensuring the proper restoration of interrupt vectors is not gilding the lily—it is essential for reliable operation.

A final word This section covers some rather complicated territory and deals with things that will not work very well (or at all) unless they are done right. If you are not already familiar with the material it might seem a little confusing (or maybe even a lot). If so, keep at it. A mastery of interrupt programming techniques can pay off handsomely in superior performance in programs you write. If you are an experienced programmer, it is hoped that this has been a helpful review.

Interrupt-driven serial I/O

Now that the general principles that underlie all hardware interrupt service routines have been discussed, it is time to apply them to interrupt-driven operation of the PC serial port.

The first decision that has to be made in planning any routines for the PC serial port is whether to use more than one of the 8250's internal "interrupt" types. If a single type is used—the RCR type, for example—the ISR will have only a single overall task to perform. If multiple types are used, the ISR must first determine which of the 8250 interrupt types caused the interrupt, and then carry out the appropriate tasks. Single-type routines are simpler and faster, but perform only one kind of function. Multiple-type ISRs can provide interrupt-driven operation of all of the 8250's functions, but they are more complicated and a little slower. As is usually the case in programming, the best choice is not absolute but depends on the application.

Interrupt service routines for the serial port can take many forms. Several examples follow. These illustrative routines have been designed for speed and efficiency. For example, flow control is not incorporated. They are based on the principle that the task of the ISR should be limited to doing nothing more than what is necessary.

Single-function serial port ISRs

Although a single-function ISR could be written for any one of the internal interrupt levels of the 8250, it would most commonly be for the receiver interrupt. Only the internal receiver interrupt in the 8250 would be enabled; bit 0 of the IER would be set to 1 and all other bits to 0.

The pseudocode in Fig. 4-7 shows the structure of an ISR for the 8250 receive interrupt. Saving the CPU registers on entry and setting up data addressing are common to all ISRs, as are sending an EOI and restoring registers at the conclusion of the ISR. The "business" part of the ISR comes between these housekeeping operations and is quite simple. The received byte is read from the data port and put in a variable in the main program's data area. This variable is usually a buffer, an array of bytes, with another variable to hold the value of the current index into the array. The ISR increments or otherwise adjusts this index variable as appropriate to reflect the newly added byte. It is useful, though not necessary, to maintain a count of the number of bytes currently in the buffer; this can be done by incrementing a count variable. The ISRs shown later for the serial toolkit give an idea of what such a routine would look like fleshed out into actual code.

A single-task ISR for the transmit function is shown in Fig. 4-8. In practice a single-purpose transmit ISR would not be used very often because there would generally be two-way data flow, but showing it as a single task helps clarify its structure.

The core of the transmit ISR is in some respects a mirror image of the receive ISR. The data to be sent is in a buffer—TxBuffer—with an index variable—TxIndex. The ISR fetches the current byte and puts it in the 8250 transmit register, and increments TxIndex to point to the next byte. To avoid going beyond the number of bytes in the buffer, a variable holding the number of bytes to be sent is decremented each time a byte is sent. When this variable reaches zero, no more bytes are transmitted. The

Fig. 4-7 *Pseudocode for a receive ISR.*

```
procedure ReceiveISR
begin
   save CPU registers
   set up data addressing
   read 8250 receive register
   put byte in buffer
   adjust buffer index
   adjust other variable(s) { e.g., count of bytes in buffer, etc. }
   send EOI to PIC
   restore CPU registers
end
```

routine could simply skip writing a byte to the 8250 transmitter if the buffer is empty, but a more elegant method is to disable the 8250 transmit interrupt. This ploy avoids the overhead of what would amount to a do-nothing interrupt being activated constantly during the time that the buffer is empty.

The pseudocode for a transmission setup routine—**SendBlock**—is also shown. The data to be sent is put into the buffer. (For compactness and efficiency an actual subroutine would receive a pointer to the buffer [a **var** parameter in Pascal] rather than loading a separate buffer structure.) *TxIndex* and *TxCount* are then initialized. Finally, the 8250 transmit interrupt is enabled and transmission begins.

The setup routine as it is shown is best suited for block-at-

Fig. 4-8 *Pseudocode for a transmit ISR.*

```
VAR { static variables }
   TxBuffer : array of bytes
   TxIndex   { points to byte to be sent }
   TxCount   { number of bytes in buffer }

procedure TransmitISR
begin
   save CPU registers
   set up data addressing
   write TxBuffer[ TxCount ] to 8250 transmit register
   increment TxIndex
   decrement TxCount
   if TxCount = 0 then
      disable IER transmit interrupt, bit 1  { IER := IER xor 2 }
   send EOI to PIC
   restore CPU registers
end

procedure SendBlock
begin
   put block to be sent into TxBuffer
   set TxCount to number of bytes to send
   set TxIndex to point to first byte in buffer
   enable 8250 transmit interrupt, IER bit 1 { IER := IER or 2 }
end
```

a-time operation, but it is possible to design a setup routine that would allow data to be added to the transmit buffer at any time. Such a modified setup routine would issue a CLI instruction to the CPU to turn off interrupts, add the data to the buffer and adjust *TxIndex* and *TxCount*, and then issue an STI instruction to allow hardware interrupt processing to resume.

Interrupt-driven transmission might, at first sight, seem attractive; once the buffer has been loaded and the 8250 transmit interrupt enabled, transmission continues automatically. The main program can be off doing other things without waiting for the transmission to be completed. This technique is indeed useful where substantial amounts of data that are inherently in block form are to be sent (a file transfer program, for example) or where the other system receiving the data requires input at a fairly steady rate. For general-purpose applications, however, transmission control by polling is usually adequate. A polling routine such as a "communication engine" (as discussed previously) will usually loop so fast that full transmission speed can be maintained. Something of a middle ground is offered by a subroutine that accepts data as blocks but transmits by polling. Such a routine is illustrated by the procedure **BlockWrSerial** in the Serial Toolkit later in this chapter.

Multiple-function serial port ISRs

When the 8250 is programmed to generate an interrupt for more than one condition, the ISR has the job of determining what caused the interrupt. The ISR must read the IIR register and then execute the appropriate code. The pseudocode for an ISR that handles all four of the 8250's interrupt types is shown in Fig. 4-9.

The receive and transmit processing sections are like the individual routines discussed previously. The remaining two sections, for line errors and modem status, simply set a flag variable in the main program to TRUE and copy the status register to a variable that can then be examined by the main program. In some cases it might be desirable to expand one or both of these status functions so that they immediately disable further transmission by setting the 8250 transmit interrupt enable bit to zero.

Summary of serial interrupt setup procedure

As you have seen, setting up an interrupt-driven serial routine and shutting it down in an orderly manner involves several nec-

Fig. 4-9 *Pseudocode for a multifunction ISR.*

```
procedure FullISR
begin
   save CPU registers
   set up data addressing
   read 8250 IIR
   if IIR = character-ready then
      read 8250 receive register
      put byte in buffer
      adjust buffer index
      adjust other variable(s) { e.g., count of bytes in buffer, etc. }
      goto Done
   if IIR = transmit-ready then
      get next byte to be sent
      write to 8250 transmit register
      adjust other variables { count of bytes sent, etc. }
      if buffer empty then disable IER transmit bit
      goto Done
   if IIR = line-error then
      read 8250 line status register
      put the byte in a line status variable
      set line error variable to TRUE
      goto Done
   if IIR = modem-status-change then
      read 8250 modem status register
      put the byte in a modem status variable
      set modem change variable to TRUE
   Done:
      send EOI to PIC
      restore CPU registers
end
```

essary steps. Although the discussion has centered specifically on the serial port, a similar process would be used for any kind of hardware interrupt. Here is a summary or checklist of the steps required:

A. Setup
   1. Save current interrupt vector and install vector to ISR.
   2. Set PIC to enable serial IRQ.
   3. Configure UART (baud, etc.).
   4. Enable UART internal interrupt level(s).
   5. Enable serial IRQ line connection.

B. Run

C. Shutdown
   6. Disable serial IRQ line connection.
   7. Restore original UART configuration (optional).
   8. Reset PIC to disable serial IRQ.
   9. Restore saved original interrupt vector.

Practical serial operation

Let's put all of these many foundational matters together and turn now to the development of a practical general-purpose serial control and I/O package. All of the steps discussed, from programming the 8250 to writing ISR code to setting up the PIC to installing and restoring interrupt vectors, are illustrated in the Serial Toolkit. It is offered for study as well as for practical use. Seeing how it is constructed should help you get a better feel for how the pieces fit together.

The first step is to decide on a general architecture. Totally interrupt-driven software is uncommon. Usually one or more key interrupt-driven service routines are combined with some form of polling. Is there a rational way to decide which operations should be polled and which should be interrupt-driven? A good way is to consider "who is in charge?" In polled operation, the program determines when to do things. With interrupts, external events call the shots. Sending should be controlled by the main program because it is usually where the data originates in the first place. Hence, polling is used for output. Receiving is, by definition, from outside. It might be (and probably will be) unpredictable, so interrupt-driven inputs are used. The combination draws on the strengths of both methods.

Although such a hybrid is very satisfactory and is the best choice for most purposes, there are special applications where other methods are more suitable. In some forms of block data I/O, for example, using interrupt-driven transmission as well as reception is the best choice. At the other end of the scale, if very slow data rates are used then pure polling without any interrupt routines might be entirely adequate and would yield smaller and simpler program code.

A second consideration is how the serial software will interface with application programs and the system. The most flexible and versatile method is to write a collection or library of service or utility subroutines, which an application program will use as a resource, most often in conjunction with some sort of "communication engine" like the one shown previously in Fig. 4-6. This is what is done in the Serial Toolkit presented later in this chapter.

It should be noted that a serial I/O package could be implemented in two other ways that are appropriate in certain cases. One is the DOS device driver. Such drivers are modules of executable code that provide the lowest-level interface between DOS and I/O devices. When DOS is started it loads any additional drivers specified in the CONFIG.SYS file, which then become part of DOS itself just like disk drives and the character devices. Another approach is the terminate and stay resident (TSR) utility. A TSR remains in memory and can be used by any program, and is typically controlled through software interrupts. For example, a high-performance serial TSR could be written that would take over interrupt vector 14h and replace the BIOS serial services. These interesting possibilities are left, as the saying goes, as exercises for the reader.

Receiving

Because only serial input (receiving) will be interrupt-driven, you enable only the character-ready internal interrupt in the UART, and use a simple single-task ISR. The job of the ISR is to put each incoming character into a buffer from which it can be retrieved later by the application program.

A data buffer is in essence an array of bytes whose elements are accessed by an index or buffer pointer. The design of a receive buffer has certain requirements that must be met in order to achieve high performance and reliability: 1) The buffer must allow the independent insertion and retrieval of bytes. This implies separate insertion (write) and retrieve (read) buffer pointers. 2) The buffer must handle overflow gracefully. In any case, the write pointer must be controlled to avoid writing to memory outside of the buffer. 3) There must be a way to tell whether the buffer is empty or still contains unread bytes. (There could, optionally, be an indication of the number of bytes in the buffer at anytime, but this is usually of no consequence.) 4) Bytes must be able to be read from the buffer in the same order in which they were put in.

The data structure that best meets all of these requirements is the ring buffer, also known as a circular queue. It is basically an ordinary array of bytes. Incoming bytes are added to the array at the current value of the buffer pointer and the pointer is incremented, just as in the usual linear array. When a byte is added to the last slot in the array, however, the buffer pointer is not incremented but instead is set to point to the first element in the array. As bytes are added, the pointer "wraps around" from tail back to head. The array has become a loop or circle, logically speaking. This satisfies the second requirement.

By adding a second buffer pointer that indicates the byte ready to be retrieved, the first and fourth requirements are satisfied. The read pointer is incremented after a byte has been read and wrapped at the end of the buffer in the same manner as the write pointer. The two pointers taken together satisfy the third requirement, the ability to detect whether the buffer is empty. Suppose you start off with the buffer completely empty and both pointers pointing to element 0. An incoming byte goes into buffer[0] and the write pointer is incremented to 1. The read pointer is still 0. Now suppose you read the buffer. You retrieve the byte at buffer[0] and increment the read pointer so it becomes 1. Notice that the write pointer and the read pointer have now become equal to each other. Because you have read the only byte that was in the buffer, it is empty once again. The same pattern holds when there are many bytes in the buffer. Whenever the write and read pointers are not equal, bytes are waiting to be read; if the pointers are equal, the buffer is empty. A little thought will show that this holds even if pointer wrap has occurred, provided that the number of yet unread bytes is less than the buffer size. If too many bytes remain unread the write pointer can come up on the read pointer from behind, and if they become equal the buffer becomes effectively empty at that moment. A ring buffer is self-flushing, as it were; overflow simply clears out the buffer and starts over—draconian with respect to loss of data, but well-behaved in never messing up anything outside of the ring buffer area.

The receive portion of the serial I/O package therefore could be based on *SerialBuffer*, an array used as a ring buffer, with two pointers *SerBufWrPtr* and *SerBufRdPtr*. The following services would be provided:

- An ISR would respond to serial interrupts by fetching the incoming byte from the UART, putting it in the buffer, and incrementing *SerBufWrPtr*, wrapping if necessary.
- Reading a byte from the serial port would actually be retrieving a byte from the buffer. You could write a function **ReadSerial** that would return the byte that *SerBufRdPtr* is pointing to and then increment the pointer, wrapping it if need be.
- You would certainly want a status routine; you could call it **RxDataWaiting** and have it return TRUE if the buffer pointers are not equal, and FALSE if they are.

Transmitting

Transmission will be handled within a polling loop similar to the "communication engine" of Fig. 4-6. For this you will need a status function—**TxReady**—to read the line status register of the UART and return TRUE if the transmitter buffer empty bit (THRE) is set, or FALSE otherwise. To transmit a byte, a simple procedure **WriteSerial()** to write a byte to the UART data port would suffice.

Flow control

The main purpose of flow control is to suspend transmission while the receiving device is not ready to accept data. Because transmission will be by polling, flow control can be made a part of the polling loop.

If the modem control lines are being used for hardware flow control, there should be a function to report the status of these lines. The most flexible method would be a function (call it **SerialStatus)** that would return the modem status register of the UART, and the flow control section of the polling loop would test the appropriate bit or bits. There should also be procedures to provide control of the modem output lines, DTR and RTS.

If embedded codes such as XON/XOFF are being used for flow control, the receive processing section of the polling loop can test for such characters and set flag variables as appropriate. An example is shown in the fragment of pseudocode in Fig. 4-10, which is the receive and transmit sections of the general polling engine (Fig.

Fig. 4-10 *Example of flow control.*

```
procedure Poll_X:
TxOk := TRUE;
while not Quit do
   begin
      if RxDataWaiting then
         begin
            get input byte B := ReadSerial
            if B = XON then TxOk := TRUE
               else if B = XOFF then TxOk := false
                  else process B
         end

      ...

      if byte-to-send then
         if TxReady and TxOk then
            get byte T, send it:  WriteSerial( T )
   end while
end Poll_X
```

4-6) modified to implement flow control. XON and XOFF characters in the incoming data stream are trapped and used to toggle the boolean variable *TxOk* to enable or disable transmission.

The serial toolkit

The Serial Toolkit, shown in Fig. 4-11, provides all of the core routines ordinarily needed for high-performance serial I/O. It implements a single serial channel through either the COM1 or COM2 serial port.

The Serial Toolkit illustrates the practical application of the principles already discussed in this chapter. It follows the "hybrid" model, combining interrupt-driven receiving with polled transmission and status checking. As with the CRC Toolkit presented in chapter 3, it is designed as a library of routines packaged in a module. The toolkit can be used as is in programs for serial I/O or modified as needed to suit your particular needs. It is also offered for your study to see how the things that have been discussed are actually done in working, tested code.

Receive buffer and the ISR

Receive (input) operations are interrupt-driven. An ISR fetches each incoming byte from the UART and puts it in a buffer. The main program gets serial-port input by retrieving bytes from the receive buffer.

The receive buffer, *SerialBuffer*, is declared explicitly and its last element (its size in bytes minus 1) is declared as a constant, *SerBufEnd*. This has the (possible) disadvantage that its size cannot be changed during execution and that setting a different size entails recompiling the unit. On the other hand, it has the advantage that it allows the ISR to be written using immediate operands, producing smaller and faster code. The size shown in the listing is 4096 bytes, which long experience has shown works well in general-purpose use from 2400 to 38,400 baud. A smaller buffer will suffice if operation will be at lower baud rates (below 4800). A larger buffer might be necessary if the main program contains time-consuming routines such as real-time graphics and rates above 9600 baud are to be used.

SerialBuffer is arranged as a ring buffer using read and write pointers *SerBufRdPtr* and *SerBufWrPtr.* A special technique is used for wrapping around the end of the buffer. The buffer size is

Fig. 4-11 *The serial toolkit unit.*

```
{ ------------------------------------------------------------------------- }
{                               SERIAL TOOLKIT                              }
{                  A Library of Serial Port I/O Routines                    }
{ ------------------------------------------------------------------------- }
(* SERIALTK.PAS 1.0 ©1992 J H Johnson *)


UNIT
   SerialTk;

INTERFACE

USES
   Dos;

CONST
   { *** SET BUFFER SIZE HERE *** }
     { serial buffer size is always 2^N }
     { last element is (2^N)-1, e.g. 255,511,1023,2047,4095,8191,16383,32767 }
   SerBufEnd = 4095;       { Must be the same as the equate in SERTKISR.ASM ! }

TYPE
   Parity = ( Npar,Epar,Opar,Mpar,Spar );          { None,Even,Odd,Mark,Space }
   UARTstate = record
                  IER,                                        { UART registers }
                  LCR,
                  MCR,
                  DivLo,                                   { divisor (as 2 bytes) }
                  DivHi : byte;
                  BA : word;                                 { port base address }
               end;


VAR
   SerCount : word;                                  { count of chrs in buffer }


{ ----- setup routines ----- }

procedure OpenSerial( P : byte );
      { P is serial port: 1 = COM1, 2 = COM2 }

procedure CloseSerial;
      { close and restore original interrupt }

procedure SetUART( Rate : word; Par : Parity; Db,Sb : byte );
      { Sets UART rate and bit values.  Rate = actual rate in bps. }
      {   [note that (115200 div Rate) must be an integer value]   }
      { Db data bits = 5..8, Sb stop bits = 1 or 2                  }

procedure EnableUART;
      { enable UART and turn on DTR, RTS, and OUT2 for ints }

procedure DisableUART;
      { disable UART and turn off DTR, RTS, OUT1, OUT2 }

procedure SaveUARTstate( var US : UARTstate );
      { save current UART state }
```

Fig. 4-11 Continued.

```
procedure RestoreUARTstate( var US : UARTstate );
      { restore current UART state }

{ ----- status routines ----- }

function SerialStatus : word;
      { returns current status:                                    }
      { bits 15..12:  CD,RI,DSR,CTS;  set = ON                     }
      { bits 11..8:   deltas: dCD,dRI,dDSR,dCTS                    }
      { bit 2 set = UART enabled                                   }
      { bits 0,1:  0=not open, 1=open as COM1, 2= open as COM2     }

function CommError : byte;
      { returns UART errors, bit significant:                        }
      {  bit 4=Break, 3=FramingErr, 2=ParityErr, 1=Overrun if set }
      {  bit 0=SerialBuffer overflow if set }

function TxReady : boolean;
      { true if OK to send a byte }

function RxDataWaiting : boolean;
      { true if Rx buffer not empty }

{ ----- processing routines ----- }

{ for AssertXXX: true = ON, false = OFF }

procedure AssertBreak( SetON : boolean );
      { when turned on, BREAK remains on until turned off }

procedure AssertDTR( SetON : boolean );

procedure AssertRTS( SetON : boolean );

procedure AssertOUT1( SetON : boolean );

procedure FlushSerialBuffer;
      { clear Rx buffer }

function ReadSerial : byte;
      { fetches next byte from Rx buffer, or 0 if buffer empty }

function PeekSerial : byte;
      { next byte in Rx buffer, 0 if buffer empty; byte remains in buffer }

procedure WriteSerial( B : byte );
      { transmit a byte }

procedure BlockWrSerial( var DataBlock; Len : word; var Sent : word );
      { BlockWrSerial sends a block of data as fast as possible; no flow }
      { control. DataBlock is array of bytes. Len is number of bytes in  }
      { DataBlock to send. Sent returns number of bytes actually sent.   }
      {    CAUTION: Len not value-checked; must be <= size of DataBlock! }

IMPLEMENTATION

{ ================================================================== }
```

Fig. 4-11 Continued.

```
{ ----- Cross-reference to Assembler declarations ----- }
{ These static variables are declared in SERTKISR.ASM:
  DATA SEGMENT BYTE PUBLIC
       EXTRN SerBufWrPtr : word
       EXTRN SerialBuffer : byte
       EXTRN SerCount : word
  DATA ENDS
  This equate MUST agree with the value in the unit: SerBufEnd equ 4095      }
{ ---------------------------------------------------- }


CONST
   { ---------- Equates for serial port driver ---------- }

   Com1base = $3F8;
   Com2base = $2F8;

   { 8250 register addresses }
   DTA = 0;                     { data I/O                    }
   IER = 1;                     { interrupt enable register }
   IIR = 2;                     { interrupt ident register  }
   LCR = 3;                     { line control register     }
   MCR = 4;                     { modem control register    }
   LSR = 5;                     { line status register      }
   MSR = 6;                     { modem status register     }

   { system hardware equates }
   Com1Int = $0C;               { IRQ4 hardware int (COM1)  }
   Int4Mask = $10;              { mask to set IRQ4          }
   Com2Int = $0B;               { IRQ3 hardware int (COM2)  }
   Int3Mask = $08;              { mask to set IRQ3          }

   { interrupt controller equates }
   IntMaskReg = $21;            { 8259 mask register port   }
   IntEOIport = $20;            { port for 8259 EOI         }
   EOI = $20;                   { 8259 EOI instruction      }


VAR
   SerialBuffer : array [0..SerBufEnd] of byte;                { Rx buffer }
   SerBufRdPtr, SerBufWrPtr : word;                 { ring buffer pointers }
   Base : word;                                       { port base address }
   Device : byte;
   IntLvl,                                              { interrupt level }
   IntMask : byte;                                { PIC mask for int level }
   SerialOn : boolean;                           { true if UART is enabled }
   OldComIntVec : pointer;                      { original interrupt vector }
   ExitSave : pointer;


{ ------------------------- private routines ----------------------------- }

{$F+ }

{$L SERTKISR.OBJ }
procedure Ser1ISR; external;
```

Fig. 4-11 Continued.

```
procedure Ser2ISR; external;

procedure SerialExit;
begin
   CloseSerial;
   ExitProc := ExitSave;
end;

{$F- }


{ -------------------------- public routines ----------------------------- }

{ ----- serial port status routines ----- }

function SerialStatus : word;
var
   W : word;
begin
   W := Device;                        { bits 0,1 : 0=not open, 1=COM1, 2=COM2 }
   if SerialOn then W := W or 4;                     { bit 2 set = UART enabled }
   hi(W) := Port[Base+MSR];     { bits 15..8: CD,RI,DSR,CTS;dCD,dRI,dDSR,dCTS }
   SerialStatus := W;
end;

function CommError : byte;
var
   B : byte;
begin
   B := Port[Base+LSR] and $1E;     { bits 4..1: Break,Framing,Parity,Overrun }
   if SerCount > SerBufEnd then B := B or 1;         { bit 0: buffer overflow }
   CommError := B;
end;

function TxReady : boolean;
begin
   TxReady := Port[Base+LSR] and $20 <> 0;
end;

function RxDataWaiting : boolean;
begin
   RxDataWaiting := SerBufRdPtr <> SerBufWrPtr;
end;

{ ---------- control routines ---------- }
{ Note: Assert routines also set bit 3 (= 08h) to ensure that OUT2 is on. }

procedure AssertBreak( SetON : boolean );
var
   B : byte;
begin
   B := Port[Base+LCR];
   if SetON then B := B or $40 else B := B and $BF; { = and not 40h }
   Port[Base+LCR] := B;
end;
```

Fig. 4-11 Continued.

```
procedure AssertDTR( SetON : boolean );
var
   B : byte;
begin
   B := Port[Base+MCR];
   if SetON then B := B or $09 { = 01h or 08h }
      else B := B and $FE; { = and not 01h }
   Port[Base+MCR] := B;
end;

procedure AssertRTS( SetON : boolean );
var
   B : byte;
begin
   B := Port[Base+MCR];
   if SetON then B := B or $0A  { = 02h or 08h }
      else B := B and $FD; { = and not 02h }
   Port[Base+MCR] := B;
end;

procedure AssertOUT1( SetON : boolean );
var
   B : byte;
begin
   B := Port[Base+MCR];
   if SetON then B := B or $0C  { = 04h or 08h }
      else B := B and $FB; { = and not 04h }
   Port[Base+MCR] := B;
end;

procedure FlushSerialBuffer;
begin
   SerBufRdPtr := 0;
   SerBufWrPtr := 0;
   SerCount := 0;
end;

{ ---------- serial port setup ---------- }

procedure OpenSerial( P : byte );
var
   Mask : word;
begin
   if (Device <> 0) or (P<1) or (P>2) then Exit;
   { set static vars for COMx, save initial int vector and set new }
     if P=2 then
        begin
           Device := 2;
           Base := Com2base;
           IntMask := Int3Mask;                          { mask to set IRQ3 }
           IntLvl := Com2Int;
           GetIntVec( IntLvl,OldComIntVec );
           SetIntVec( IntLvl,@Ser2ISR );               { set up int for COM2 }
        end else begin
```

Fig. 4-11 Continued.

```
            Device := 1;
            Base := Com1base;
            IntMask := Int4Mask;                          { mask to set IRQ4 }
            IntLvl := Com1Int;
            GetIntVec( IntLvl,OldComIntVec );
            SetIntVec( IntLvl,@Ser1ISR );                 { set up int for COM1 }
         end;
   { set up hardware interrupt }
      Mask := Port[IntMaskReg];
      Port[IntMaskReg] := Mask and (not IntMask);      { zero the relevant bit }
   FlushSerialBuffer;
end;

procedure CloseSerial;
var
   Mask : word;

begin
   if Device = 0 then Exit;
   Port[Base+MCR] := Port[Base+MCR] xor $08;        { OUT2 off to disable ints }
   Mask := Port[IntMaskReg];                       { reset interrupt controller }
   Port[IntMaskReg] := Mask or IntMask;                  { set the relevant bit }
   SetIntVec( IntLvl,OldComIntVec );             { restore orig interrupt vector }
   Device := 0;
end;

procedure SetUART( Rate : word; Par : Parity; Db,Sb : byte );
const
   ParMask : array [Npar..Spar] of byte = ( 0,$18,$08,$28,$38 );
var
   R : word;
begin
   if (Device = 0) or (Rate < 2) then Exit
      else R := 115200 div Rate;                      { calculate rate divisor }
   Db := (Db-5) and 3;                                      { nr of data bits }
   if Sb = 2 then Db := Db or 4;                            { nr of stop bits }
   Db := Db or ParMask[Par];                                { set parity bits }
   Port[Base+MCR] := $10;                   { loopback (req'd by early 8250s) }
   Port[Base+LCR] := $80;                                     { set DLAB = 1 }
   Port[Base] := lo( R );                              { load rate divisor LSB }
   Port[Base+IER] := hi( R );                              { load divisor MSB }
   Port[Base+LCR] := Db;                              { set bits with DLAB = 0 }
   Port[Base+MCR] := 0;                                      { leave disabled }
end;


procedure EnableUART;
var
   B : byte;
begin
   if Device = 0 then Exit;
   B := Port[Base];                                  { clear out any stray char }
```

Fig. 4-11 Continued.

```
   B := Port[Base+LSR];                            { read LSR to clear it }
   B := Port[Base+MSR];                                     { ditto MSR }
   Port[Base+IER] := 1;                   { enable 8250 Rx interrupt only }
   Port[Base+MCR] := $B;                     { turn on RTS & DTR & OUT2 }
   SerialOn := true;
end;

procedure DisableUART;
begin
   if Device = 0 then Exit;
   Port[Base+IER] := 0;                     { turn off 8250 interrupt line }
   Port[Base+MCR] := 0;                   { turn off DTR, RTS, OUT1, OUT2 }
   SerialOn := false;
end;

procedure SaveUARTstate( var US : UARTstate );
begin
   if Device = 0 then Exit;
   US.IER := Port[Base+IER];                            { save registers }
   US.LCR := Port[Base+LCR];
   US.MCR := Port[Base+MCR];
   Port[Base+MCR] := 0;                         { disable UART interrupts }
   Port[Base+LCR] := $80;                                      { set DLAB }
   US.DivLo := Port[Base];                                 { save divisor }
   US.DivHi := Port[Base+IER];
   Port[Base+LCR] := US.LCR;                 { restore changed registers }
   Port[Base+MCR] := US.MCR;
   US.BA := Base;
end;

procedure RestoreUARTstate( var US : UARTstate );
begin
   Port[US.BA+MCR] := 0;                        { disable UART interrupts }
   Port[US.BA+LCR] := $80;                                     { set DLAB }
   Port[US.BA] := US.DivLo;                             { restore divisor }
   Port[US.BA+IER] := US.DivHi;
   Port[US.BA+LCR] := US.LCR;                         { restore registers }
   Port[US.BA+IER] := US.IER;
   Port[US.BA+MCR] := US.MCR;
end;

{ ---------- I/O routines ---------- }

function ReadSerial : byte;
begin
   if SerBufRdPtr <> SerBufWrPtr then
      begin
         Inline($FA);                                   { CLI ; ints off }
         ReadSerial := SerialBuffer[SerBufRdPtr];
         SerBufRdPtr := succ(SerBufRdPtr) and SerBufEnd;
         Inline($FB);                               { STI ; ints back on }
         dec(SerCount);
      end else ReadSerial := 0;
end;
```

Fig. 4-11 Continued.

```
function PeekSerial : byte;
begin
   if SerBufRdPtr <> SerBufWrPtr then
      PeekSerial := SerialBuffer[SerBufRdPtr]
         else PeekSerial := 0;
end;

procedure WriteSerial( B : byte );
begin
   Port[Base] := B;
end;

{ BlockWrSerial sends a block of data as fast as possible; no flow control. }
{ DataBlock is array of bytes. Len is number of bytes in DataBlock to send. }
{ Sent returns number of bytes actually sent.                               }
{    CAUTION: Len not value-checked; must be <= size of DataBlock!          }

procedure BlockWrSerial( var DataBlock; Len : word; var Sent : word );
type
   ByteArr = array [1..MaxInt] of byte;          { max block size 32767 bytes }
var
   K : word;
   DB : ByteArr absolute DataBlock;
   TO : longint;
   Ticks : longint absolute $40:$006C;                   { 18.2 Hz event clock }
begin
   Sent := 0;
   for K := 1 to Len do
      begin
         TO := Ticks;
         while not (Port[Base+LSR] and $20 <> 0) do         { wait for tx ready }
            if Ticks-TO > 4 then Exit;               { exit on timeout, 200 ms }
         Port[Base] := DB[K];
         inc( Sent );
      end;
end;

{ ----------------------------- initialization ----------------------------- }

BEGIN
   ExitSave := ExitProc;
   ExitProc := @SerialExit;
   Base := Com1base;
   Device := 0;
   SerialOn := false;
END.
```

constrained to be a power of 2 so that it consists of 2^N elements, 0 to $2^N - 1$. Because $2^N - 1$ will always be all 1s, the pointer can be wrapped simply by ANDing it with $2^N - 1$. This is very much faster than comparing the pointer to a maximum value and setting it to zero if it is greater.

If you want to use a different receive buffer size, remember that the fast wrap ploy permits only certain sizes. The value specified for *SerBufEnd* must be some $2^N - 1$; for instance, 63, 127, 255, 511, 1023, 2047, 4095, 8191, 16,383, or 32,767. (The corresponding hexadecimal values are 3F, 7F, FF, 1FF, 3FF, 7FF, FFF, 1FFF, 3FFF, 7FFF.) **Important:** If you change buffer size and you are using the assembly language ISR, you must change the *SerBufEnd* equate in the .ASM file also!

The ISR (Fig. 4-12) is written in assembly language for compactness and speed. The assembler code is written in "classic" Intel-Microsoft form that should work with any 80x86 assembler. (Borland TASM was used for the examples in this book.) In Turbo Pascal 4.0 and later, assembly language code is linked by specifying the object-language (.OBJ) file with the $L compiler directive. As is usual in Pascal, such routines must be marked as **external.** Linking with other languages is generally along similar lines, though segment names and possibly some identifiers might have to be changed. Most C compilers, for example, require external public identifiers to have a leading underscore. The medium memory model is assumed, with multiple code segments and a single static data segment.

An alternate version of the ISRs written in Pascal is shown in Fig. 4-13. You can use it rather than the assembler version with a slight sacrifice of speed. You might also find it helpful to compare the Pascal code with the assembler version to see just what is going on; both versions work in the same way. Whichever version you use, don't forget that these are far (segment:offset) routines and require the $F+ compiler directive.

There are actually two ISRs, one for COM1 and one for COM2. This allows the full use of immediate operands—constant values put into registers directly, rather than fetched from memory—resulting in smaller and faster code. Having two versions of the same routine really isn't very wasteful because the machine language for each ISR is only 43 bytes long.

The ISR follows the principles discussed earlier in this chapter. First, all CPU registers that will be modified are saved on the stack. Next, the address of the data segment is obtained (it is plugged in when DOS loads the program) so that the data segment variables *SerialBuffer, SerBufWrPtr*, and *SerCount* can be accessed. The incoming byte is fetched from the UART with an INput instruction and put in the buffer. The write pointer *SerBufWrPtr* is incremented and ANDed with *SerBufEnd*, which wraps the pointer if necessary. The variable *SerCount* is incremented (more on that in a moment). This completes the work of the ISR, so an end-of-interrupt command is sent to the interrupt

Fig. 4-12 *The serial toolkit ISRs.*

```
;--------------------------------------------------------------
; Interrupt Service Routines for serial ports COM1 and COM2
;        COM1: Ser1ISR       COM2: Ser2ISR
;--------------------------------------------------------------
; SERTKISR.ASM 2.1 ©1992 J H Johnson

DATA    SEGMENT BYTE PUBLIC

        EXTRN SerBufWrPtr : word
        EXTRN SerialBuffer : byte
        EXTRN SerCount : word

DATA ENDS

CODE    SEGMENT BYTE PUBLIC

        assume cs:code,ds:data

; buffer size equate -- must match constant declared in unit
        SerBufEnd        equ 4095

; general equates
        Com1_data        equ 3F8h
        Com2_data        equ 2F8h
        IntEOIport       equ 20h
        EOI              equ 20h


        PUBLIC Ser1ISR

Ser1ISR PROC FAR

        push    ax                              ; save all the registers
        push    dx                              ; that will be used
        push    di
        push    ds

        mov     ax,SEG DATA                     ; set up data seg addr
        mov     ds,ax
        mov     dx,Com1_data                    ; COM1 port addr for data

        in      al,dx                           ; get byte from port

        mov     di,[SerBufWrPtr]                ; get buffer write pointer
        mov     [SerialBuffer+di],al            ; store byte therein
        inc     di                              ; increment write pointer
        and     di,SerBufEnd                    ; wrap if necessary
        mov     [SerBufWrPtr],di                ; put new value in variable
        inc     [SerCount]                      ; increment SerCount
        mov     al,EOI                          ; set PIC port addr
        out     IntEOIport,al                   ; EOI cmd to PIC

        pop     ds                              ; restore registers
        pop     di
        pop     dx
        pop     ax
        iret                                    ; return from interrupt

Ser1ISR ENDP
```

Fig. 4-12 Continued.

```
;--------------------------------------------------------------

        PUBLIC Ser2ISR

Ser2ISR PROC FAR

        push    ax
        push    dx
        push    di
        push    ds

        mov     ax,SEG DATA
        mov     ds,ax
        mov     dx,Com2_data                        ; COM2 port

        in      al,dx

        mov     di,[SerBufWrPtr]
        mov     [SerialBuffer+di],al
        inc     di
        and     di,SerBufEnd
        mov     [SerBufWrPtr],di
        inc     [SerCount]

        mov     al,EOI
        out     IntEOIport,al

        pop     ds
        pop     di
        pop     dx
        pop     ax
        iret

Ser2ISR ENDP


CODE    ENDS

END
```

Fig. 4-13 *Pascal version of the serial toolkit ISRs.*

```
{ Pascal version of ISRs in the Serial Toolkit }

CONST
   Com1data = $3F8;
   Com2data = $2F8;

{ These routines must be declared as FAR }

{$F+ }

procedure Ser1ISR;
interrupt;
begin
```

Fig. 4-13 Continued.

```
   SerialBuffer[SerBufWrPtr] := Port[Com1data];                { get the char }
   SerBufWrPtr := succ(SerBufWrPtr) and SerBufEnd;            { bump ptr; wrap }
   inc(SerCount);                                      { incr buffer chr count }
   Port[IntEOIport] := EOI;                           { send EOI to controller }
end;

procedure Ser2ISR;
interrupt;
begin
   SerialBuffer[SerBufWrPtr] := Port[Com2data];                { get the char }
   SerBufWrPtr := succ(SerBufWrPtr) and SerBufEnd;            { bump ptr; wrap }
   inc(SerCount);                                      { incr buffer chr count }
   Port[IntEOIport] := EOI;                           { send EOI to controller }
end;

{$F- }
```

controller, the saved CPU registers are restored, and an interrupt return (IRET) instruction returns control to whatever was executing when the interrupt occurred.

The *SerCount* variable is incremented each time the ISR executes, and it is decremented each time a byte is retrieved from the buffer. *SerCount* is included for two reasons. One is to allow an overflow of *SerialBuffer* to be detected. Ordinarily the long-term average value of *SerCount* will be zero because each increment of its value by the ISR is offset by a read from the buffer. If the buffer overflows, the reads start over again at zero so *SerCount* will not be fully decremented. This condition is reported by bit 0 in the byte returned by the *CommError* function. The other purpose of *SerCount* is to allow buffer usage to be monitored. If its value is read immediately after a time-consuming routine such as writing to a floppy disk you can get an idea of how many bytes have accumulated in the meantime. This can be helpful in selecting a reasonable buffer size (though too big is always better than too small).

Control and communication parameter functions

OpenSerial() and **CloseSerial** open and close the serial I/O channel, much as you would open or close a disk file. They encapsulate the installation and removal of the ISR and IRQ. **OpenSerial** is called with the parameter 1 or 2 to select COM1 or COM2, respectively, to be the active port. **CloseSerial** restores the original interrupt and must be called at some point before the main program terminates and exits to DOS. (As a precaution, the exit routine of the unit calls it.)

SetUART() configures the UART for baud rate, parity, and the number of data and stop bits. The baud rate parameter is the desired rate in bits per second, and can be any value from 2 to 57,600. Bear in mind that the UART clock divider is set to (115,200 div Rate), and if 115,200 is not exactly divisible by Rate the actual rate will be rounded up to the next highest integral rate. Parity is specified by one of the declared constants *Npar*, no parity; *Epar*, even parity; *Opar*, odd parity; *Mpar*, mark parity; and Spar, space parity. The number of data bits parameter must be in the range five to eight, and the number of stop bits must be one or two.

EnableUART must be called after the serial channel is opened and the UART is set in order to activate the channel. It turns on the PC IRQ interrupt, clears the UART data and status registers, and turns on the DTR and RTS lines. **DisableUART** disconnects the serial port interrupt line and turns off the DTR, RTS, and OUT1, but otherwise leaves the serial channel open. **EnableUART** and **DisableUART** can be called freely as needed by a program to suspend or resume serial channel activity.

Here is an example of how to open a serial channel. The setup routines work together and should be called in sequence. For example, to open the channel using the COM1 serial port and operating with 2400 baud, no parity, eight data bits, and one stop bit, issue the following calls:

```
OpenSerial( 1 );
SetUART( 2400, Npar, 8, 1 );
EnableUART;
```

FlushSerialBuffer clears the serial buffer. It is a convenient way to discard unwanted information or garbage. It is a good idea to call it after opening the serial channel to make sure that the buffer is empty.

The **AssertXXX** procedures control the modem and output lines **(AssertDTR, AssertRTS, AssertOUT1)** and the break state **(AssertBreak).** Calling the procedure with a TRUE parameter turns it on, and likewise a FALSE parameter turns it off. Note that if the break state is turned on it will remain in effect until it is explicitly turned off.

SaveUARTstate() reads the current state of the UART registers and the port base address into a record of type **UARTstate** that is passed to it. **RestoreUARTstate()** reverses the process and sets the UART to the values passed to it in a record of type **UARTstate.** These two procedures can be used to reset the UART at the conclusion of a program to the communication parameters that it started with. Because the values are passed in a record variable declared in the main program, you can create several such variables and use them to switch between several UART setups. **Note:** the serial channel must have been opened by **OpenSerial** before calling **SetUARTstate.**

Status functions

RxDataWaiting returns TRUE if the receive buffer is not empty; that is, there are one or more characters in the receive buffer waiting to be read. **TxReady** returns TRUE if the UART is ready to accept the next byte for transmission.

SerialStatus returns a sixteen-bit word (not a byte) whose bits reflect the current serial channel status as follows:

1. Upper byte (returns UART modem status register):
   - bits 15–12: CD, RI, DSR, and CTS lines; bit set = ON.
   - bits 11–8: delta CD, RI, DSR, CTS; bit set = line has changed state.
2. Lower byte (returns serial channel status):
   - bit 2: set if UART is enabled; clear if UART disabled.
   - bits 0,1: serial channel open status, where 00 = not open; 01 = open as COM1; and 10 = open as COM2.

CommError returns 0 if there are currently no receive errors. If one or more bits are set, the following errors exist:

- bit 4: break
- bit 3: framing error
- bit 2: parity error (if parity enabled)
- bit 1: UART receive overrun
- bit 0: receive buffer overflow

Note: bits 4 through 1 report the UART line status register.

I/O functions

ReadSerial reads and removes a byte from the receive buffer. If the buffer is empty it returns 0.

PeekSerial reads a byte from the buffer but does not remove it. If the buffer is empty it returns 0. It provides a one-byte look ahead that is sometimes useful.

WriteSerial sends a byte to the UART transmitter input register for transmission. **TxReady** should be checked beforehand to make sure the UART is ready for the next byte.

BlockWrSerial transmits a block of data (an array of contiguous bytes) rather than a single byte. It is similar in function to the fast **BlockWrite** procedure in Turbo Pascal and **_write** in Turbo C, which are direct interfaces to DOS function 40h. The *DataBlock* parameter is the block to write, and *Len* is the exact length of the block in bytes. *Sent* returns the number of bytes actually sent. If *Sent* is not equal to *Len* after **BlockWrSerial** is called, an error (presumably a time-out) occurred. There is no provision for flow control; if it is required, **WriteSerial** should be used instead.

Essentially **BlockWrSerial** is a loop that gets each byte in succession from *DataBlock*, waits until the UART is ready, and puts the byte in the UART transmit buffer. The routine borrows the system "tick" to implement a simple timer. The tick is a long (thirty-two-bit) integer located in the BIOS data area at 40:006C hex; it is incremented by one of the system counter timers every 54 ms. If the UART does not become ready for another byte within four ticks, about 200 ms, the routine quits looping and returns. (The time-out period can be increased, but 200 ms is adequate for everything above 5 baud.) This is an example of "defensive programming:" try to imagine anything that could go wrong, like a bum UART, and provide ways to back out more or less gracefully.

As with all **var** parameters in Turbo Pascal, the parameter *DataBlock* is just a far pointer. Because no type is declared for *DataBlock*, the type *ByteArr* and the dummy variable *DB* are used to inform the compiler that the untyped parameter is to be treated like an array. No memory is allocated for these items. Less strongly typed languages such as C permit untyped parameters without such dummy typing.

This is a good place to say something about *transmission efficiency*, the ratio of the actual to the ideal transmission rate. A test of **BlockWrSerial** using 222-byte blocks at 19,200 baud, for which the ideal rate is 1920 bytes per second, gave an average transmission rate on an 8-MHz XT of 1915.32 b/s, a transmission efficiency of 99.76 percent.

Extensions and modification to the serial toolkit

The Serial Toolkit and its constituent routines can be extended and modified in many ways. An obvious and useful extension would be a provision for using both COM1 and COM2 simultaneously. A second buffer with its associated pointers would be needed, as would a variable in which to save the second original interrupt. Two separate ISRs are already provided. The **OpenSerial** routine would be split into two separate procedures and a second **CloseSerial** routine written. The various supporting routines would have to be modified so that they could be directed to the appropriate port addresses. This could be done by adding a parameter to each routine to select between ports; in most cases, the port selection parameter would simply set *Base* to 3F8h for COM1 or 2F8h for COM2.

Another modification that might be useful in some cases would be interrupt-driven transmission. A transmit buffer would have to be added. Routines would be provided to load the transmitter buffer, either a byte at a time or by copying multiple bytes to it. The best way to set up the transmit buffer would be as a ring buffer. Putting bytes to be sent into the buffer would move the write pointer, and each byte transmitted would move the read pointer. The sample transmit ISR shown earlier in Fig. 4-8 illustrates how the actual transmission routines could be set up.

Applications of the serial toolkit

The best way to see how the principles and elements fit together is to examine a complete working example. The stock illustration for serial port routines seems to be the video terminal emulator. There are some good reasons for this, actually; such a program touches nearly all the bases in one way or another. A terminal is inherently bidirectional, and the program must handle transmission, reception, keyboard input, and screen management all at the same time. For these reasons, TinyTerm, a simple terminal emulator program, is presented here for your study and use. The listing is shown in Fig. 4-14. (Chapter 7 presents a sample program specifically for data acquisition.) The services provided by the Serial Toolkit give TinyTerm superior transmission performance, and it handles fast data rates with ease (most of the testing was done at 9600 or 19,200 baud).

Fig. 4-14 *Example of a simple terminal emulation program.*

```
{ ------------------------------------------------------------------------- }
{                                  TINYTERM                                 }
{               Simple Terminal Emulation Demonstration Program             }
{ ------------------------------------------------------------------------- }
TERMDEMO.PAS 1.0 ©1992 J H Johnson


USES
   Crt,SerialTk;

CONST
{ control character equates }
   ESC = #27;
   CR = ^M;
   LF = ^J;
   XON = ^Q;
   XOFF = ^S;
{ constants and parameters }
   TxBufEnd = 255;                                          { must be 2^N-1 }
   TermID : string[8] = #27'[?1c';                 { VT100 identification }
   NormColor : byte = 14; { yellow }
   BoldColor : byte = 15; { white }
   RevColor : byte = $1E; { yellow on blue }

VAR
{ transmit buffer }
   TxBuf : array [0..TxBufEnd] of char;       { ring buf of chars to transmit }
   TxWrPtr,TxRdPtr : byte;                          { transmit buffer pointers }
{ escape sequence buffer }
   EscBuf : string;                            { accumulates escape sequences }
   EscPos : byte;                                        { position in EscBuf }
   EscParam : array [1..16] of integer;           { queue of ANSI parameters }
   EscParamPos,
   NrEscParams : byte;
{ global static variables }
   RxCh, KbCh : char;
   TxGo, InEsc, Echo, Quit : boolean;
   Error : integer;

{ ----- communication routines ----- }

{ These routines to add a char or a string to Tx buffer are purposely simple }
{ for clarity; there are better ways. Note buffer overflow is not reported.  }

procedure SendChar( C : char );                     { add char C to Tx buffer }
begin
   TxBuf[TxWrPtr] := C;
   TxWrPtr := (TxWrPtr+1) and TxBufEnd;
end;
```


```
procedure SendChar( C : char );                     { add char C to Tx buffer }
begin
   TxBuf[TxWrPtr] := C;
   TxWrPtr := (TxWrPtr+1) and TxBufEnd;
end;
```


```
procedure SendString( S : string );               { add string S to Tx buffer }
var
```

Fig. 4-14 Continued.

```
   P : byte;
begin
   for P := 1 to length(S) do SendChar( S[P] );
end;

procedure FlushTxBuffer;
begin
   TxWrPtr := 0;
   TxRdPtr := 0;
end;

{ ----- some simple screen routines using Turbo CRT library ----- }

procedure MoveCursor( C : char; N : byte );
begin
   case C of
      'H' : GotoXY( 1,1 );
      'U' : GotoXY( whereX, whereY-N );
      'D' : GotoXY( whereX, whereY+N );
      'L' : GotoXY( whereX-N, whereY );
      'R' : GotoXY( whereX+N, whereY );
      end; {case}
end;

procedure SGR( N : integer );
begin
   case N of
      0,2 : TextAttr := NormColor; { normal }
      1 : TextAttr := BoldColor; { bold }
      5 : TextAttr := TextAttr + 128; { blink }
      7 : TextAttr := RevColor; { reverse video }
      end; {case}
end;

procedure CPR;                                  { send current cursor position }
var
   S : string[8];
begin
   str( whereY:0,S );
   SendString( #27'[' + S + ';' );
   str( whereX:0,S );
   SendString( S + 'R' );
end;

procedure ModeSet( N : integer );
begin
   { set modes such as wrap, auto line feed, etc. }
end;

procedure ModeReset( N : integer );
begin
   { reset modes such as wrap, auto line feed, etc. }
end;

{ ----- terminal response routines ----- }

procedure SendCursor( Ch : char );
begin
   case Ch of
      'H' : SendString( #27'[H' );
      'U' : SendString( #27'[A' );
      'D' : SendString( #27'[B' );
      'R' : SendString( #27'[C' );
      'L' : SendString( #27'[D' );
```

Fig. 4-14 Continued.

```
            end; {case}
         if Echo then MoveCursor( Ch,1 );
      end;

      procedure StatusReport( N : integer );
      begin
         case N of
            5 : SendString( #27'[0n' );               { status response "ok" }
            6 : CPR;                                   { report cursor pos }
            end;
      end;

      procedure Identify;
      begin
         SendString( TermID );
      end;

      { ----- ANSI command parser ----- }

      function NextArg( N : integer ) : integer;      { N is default value }
      begin
         if EscParamPos < NrEscParams then
            begin
               inc( EscParamPos );
               NextArg := EscParam[EscParamPos];
            end else NextArg := N;
      end;

      procedure Arg;            { get a numerical parameter and store as integer }
      var
         S : string[8];
         N,E : integer;
      begin
         if EscBuf[EscPos] in ['0'..'9'] then
            begin

            S := '';
            repeat
               S := S + EscBuf[EscPos];
               inc( EscPos );
            until not (EscBuf[EscPos] in ['0'..'9']) or (EscPos > length(EscBuf));
            val( S,N,E );                         { convert string to integer }
            if E=0 then
               begin
                  inc( NrEscParams );
                  EscParam[NrEscParams] := N;       { add to parameter queue }
               end;
         end;
      if EscBuf[EscPos] = ';' then
         begin
            inc( EscPos );
            Arg;                          { Arg recurses to get all parameters }
         end;
   end;

   procedure Execute;
   var
      Temp : integer;
   begin
      Arg;                                        { numerical parameter }
      case EscBuf[EscPos] of
```

Fig. 4-14 Continued.

```
      'A' : MoveCursor( 'U',NextArg(1) );            { relative cursor moves }
      'B' : MoveCursor( 'D',NextArg(1) );
      'C' : MoveCursor( 'R',NextArg(1) );
      'D' : MoveCursor( 'L',NextArg(1) );
      'H','f' : begin                                 { absolute cursor move }
              Temp := NextArg(1) ;                { must swap; ANSI is row,col }
              GotoXY( NextArg(1),Temp );
           end;
      'J' : if NextArg(0) = 2 then ClrScr;
      'K' : begin
              Temp := NextArg(0);
              case Temp of
                 0 : ClrEol;
                 2 : begin write( CR ); ClrEol; end;           { clear line }
                 end; {case}
           end;
      'L' : for Temp := 1 to NextArg(1) do InsLine; { use Turbo routines for }
      'M' : for Temp := 1 to NextArg(1) do DelLine;    { insert, delete line }
      'c' : Identify;
      'h' : repeat ModeSet( NextArg(0) ) until EscParamPos = NrEscParams;
      'l' : repeat ModeReset( NextArg(0) ) until EscParamPos = NrEscParams;
      'm' : repeat SGR( NextArg(0) ) until EscParamPos = NrEscParams;
      'n' : StatusReport( NextArg(0) );
      end; {case}
end;

 procedure Private;
 begin
    inc( EscPos );
    { Private sequences, such as DEC "?" codes, would be handled here. }
 end;

 procedure Compound;
 var
    Temp : integer;
 begin
    inc( EscPos );
    if EscBuf[EscPos] in [':'..'?'] then Private;
    Execute;
 end;

 procedure ParseEscBuf;
 begin
    EscPos := 1;
    EscParamPos := 0;
    NrEscParams := 0;
    case EscBuf[EscPos] of
       '[' : Compound;
       'D' : write( LF );                                        { Line Feed }
       'M' : begin                                       { Reverse Line Feed }
               if whereY = 1 then InsLine;
               MoveCursor( 'U',1 );
            end;
       'Z' : Identify;
       { else parse error }
       end; {case}
 end;

 { ----- event handlers ----- }
```

Fig. 4-14 Continued.

```
procedure ProcessEsc;
begin
   EscBuf := EscBuf + RxCh;
   if RxCh in ['@'..'Z','\'..'~'] then                { a final character }
      begin
         ParseEscBuf;
         InEsc := FALSE;
         EscBuf := '';
      end;
end;

procedure ProcessRx;              { read serial buffer and dispatch character }
begin
   RxCh := chr( ReadSerial );
   if RxCh = ESC then InEsc := not InEsc else
      begin

         if InEsc then ProcessEsc else write( RxCh );
      end;
end;

procedure Transmit;                             { send next char in Tx buffer }
begin
   if TxReady then
      begin
         WriteSerial( byte( TxBuf[TxRdPtr] ) );
         TxRdPtr := (TxRdPtr+1) and TxBufEnd;
      end;
end;


procedure ProcessKbd;                                        { read keyboard }
begin
   KbCh := ReadKey;
{ To use flow control, remove the comment symbols from the next 2 lines. }
   (* if KbCh = XOFF then TxGo := FALSE; *)               { for flow control }
   (* if KbCh = XON then TxGo := TRUE; *)
   if KbCh = #0 then                                      { "extended" keys }
      begin
         case ReadKey of
            { Alt- keys }
            #16 : Quit := TRUE;          { Alt-Q }
            #18 : Echo := not Echo;      { Alt-E }
            { arrow keys }
            #71 : SendCursor( 'H' );    { home }
            #72 : SendCursor( 'U' );    { curs up }
            #75 : SendCursor( 'L' );    { curs L }
            #77 : SendCursor( 'R' );    { curs R }
            #80 : SendCursor( 'D' );    { curs dn }
            { function keys }
            #59 : SendString( #27'OP' );   { F1 }
            #60 : SendString( #27'OQ' );   { F2 }
            #61 : SendString( #27'OR' );   { F3 }
            #62 : SendString( #27'OS' );   { F4 }
            end; {case}
      end else
      begin                                                  { normal key }
         if Echo then write( KbCh );
         SendChar( KbCh );
      end;
end;

procedure ProcessError;
begin
   { routines to handle errors go here }
end;
```

Fig. 4-14 Continued.

```
{ Dispatch is the core of the program's operations. }

procedure Dispatch;
begin
   { initialize }
   Quit := false;
   Error := 0;
   TxGo := TRUE;
   Echo := TRUE;
   InEsc := FALSE;
   EscBuf := '';
   FlushTxBuffer;
   FlushSerialBuffer;
   { run loop }
   repeat
      if KeyPressed then ProcessKbd;
      if RxDataWaiting then ProcessRx;
      if (Error <> 0) then ProcessError;
      if (TxRdPtr <> TxWrPtr) and TxGo then Transmit;
   until Quit;
end;

{ ----- setup ----- }

procedure SetPort;
var
   P : byte;
begin
   write( 'Serial Port: 1 for COM1, 2 for COM2, or 0 to exit: ');
   readln( P );
   if (P=1) or (P=2) then OpenSerial( P ) else Halt(0);
end;

procedure SetComm;
var
   R : word;
   DB,SB : byte;
   C : char;
   P : Parity;
begin
   write( 'Baud rate: ');  readln( R );
   write( 'Parity: [None  Even  Odd] ');
      C := UpCase( ReadKey );
      writeln( C );
   write( 'Data bits: ');  readln( DB );
   write( 'Stop bits: ');  readln( SB );


   case C of
      'E' : P := Epar;
      'O' : P := Opar;
      else  P := Npar;
      end; {case}
   SetUART( R,P,DB,SB );
end;

procedure SetColors;
begin
```

Fig. 4-14 Continued.

```
   if LastMode = 7 then                      { reset colors for monochrome }
      begin
         NormColor := 7;
         RevColor := $70;
      end;
end;

BEGIN
   SetColors;
   TextAttr := NormColor;
   ClrScr;
   writeln( 'TINY TERMINAL DEMONSTRATION' );
   writeln;
{ set up }
   SetPort;
   SetComm;
   writeln;
   writeln( 'Alt-E toggles local echo.   Alt-Q to quit and exit.');
   write( 'Press a key to begin. ' );
   if ReadKey > #0 then ;
{ go }
   ClrScr;
   EnableUART;
   Dispatch;
{ close down }
   CloseSerial;
END.
```

The simplest terminal is the "dumb" or straight TTY terminal that sends from the keyboard and displays input characters on the screen—the "glass teletype." Just about all terminals nowadays are "smart" and can respond to escape sequences (control codes) to move the cursor, change the screen, and, in the most elaborate types, display graphics. The usual terminal emulator sample program implements a terminal that is smart, but not very. Only a few basic functions are provided and the escape sequences are simple ones used in obsolete equipment or popular and good but obsolescent terminals such as the Digital Equipment (DEC) VT52. Sample programs of this sort are interesting, but hardly practical. The fact is that the ANSI X3.64 standard escape sequences are now virtually standard. Every sample program I have seen begs off from ANSI sequences, saying they are "too complicated" for a simple demonstration program.

Well, that is a challenge. It also goes against the grain of this book, which tries to present practical (that is, actually usable) routines and examples. For that reason, TinyTerm is a very modest but genuine ANSI terminal recognizing a subset of the escape sequences used in such ANSI terminals as the DEC VT100 series.

It even identifies itself to DEC computer systems as a basic VT100. TinyTerm implements a limited but useful set of ANSI commands:

- Relative cursor movement (*N* times).
- Absolute cursor positioning to row, column.
- Clear to end of line; clear line; clear screen.
- Insert and delete line.
- Normal, bold, blink, and reverse video attributes.
- Cursor position report.
- Up, down, left, right, and home cursor ("arrow") keys.
- DEC function keys PF1–PF4.

TinyTerm uses the Serial Toolkit for serial port services. The interrupt-driven receive provides reliable operation at high speeds (for example, 38,400 baud). A ring buffer is set up for transmission. This is useful even though transmission is polled, because strings to be sent can be put in it (alternatively, **Block-WrSerial** could be used to send strings, but it would not permit character-by-character flow control to be used).

TinyTerm draws on the Turbo Pascal Crt Unit library for screen management. The unit provides a number of useful routines for text-mode screen control and writes directly to the screen buffer for speed and flexibility. Many language compilers offer similar libraries. (If you are using a language that doesn't have similar routines you can use the video BIOS [interrupt 10h] for these operations.) The procedure **GoToXY(*x,y*)** positions the cursor at column *x* and row *y*. **WhereX** and **WhereY** return the current cursor column and row. The home position in all of these routines is defined as (1,1); the BIOS and some other libraries might define it as (0,0). **InsLine** and **DelLine** insert and delete a screen row respectively at the current cursor *y* location. **ClrScr** clears the entire screen. The public variable *TextAttr* is a byte that holds the current character attribute; it is passed directly to the screen buffer during character writes. The public variable *LastMode* is a word that holds the current video mode in the lower byte; the upper byte is 1 for extended screen modes (forty-three or fifty lines) with EGA and VGA video adapters. The modes are the same as in the video BIOS; mode 3 is color text and mode 7 is the monochrome adapter (MDA). As part of its initialization sequence, the Crt Unit gets the current video mode from the BIOS. TinyTerm checks *LastMode* during start-up and if the MDA was detected resets the default character colors for monochrome.

The core of the program is the **Dispatch** procedure, a communication engine that constantly checks the status of the principal program states and calls other routines as appropriate. **ProcessRx** handles incoming characters, **Transmit** sends the next character from the transmit buffer, **ProcessKbd** deals with keyboard input, and **ProcessError** handles receive and other errors. You will see in a moment how each of these functional areas is implemented. For simplicity and clarity—this is a sample program meant to illustrate principles—a number of routines such as **ProcessError** are left empty or incomplete. They can readily be fleshed out.

Receive

Incoming characters are held in the toolkit buffer until needed. As each character is fetched by **ProcessRx** from the buffer it is checked to see if it is a special character. If it is ESC it signals the start of an escape sequence, and further characters are diverted to a small auxiliary buffer dedicated to escape sequences. ANSI escape sequences end in a character between @ and ~ (but not [). When one of the ANSI terminals is detected, the escape sequence buffer is closed and passed on to the command parser. Other special characters can be trapped here; the sample program shows how XON and XOFF could be used to control a flag for flow control.

Transmit

A small ring buffer is used to hold characters waiting to be sent. Program routines do not transmit directly but put characters into the transmitter buffer as a single character **(SendChar)** or a string **(SendString)**. The transmit buffer status is checked by the **Dispatch** loop and if there are any characters in the buffer the next one is sent by the procedure **Transmit**. A flag variable *TxGo* is provided to suspend transmission for flow control or otherwise. The sample program does not implement flow control so *TxGo* is always TRUE.

Keyboard

The dispatch loop checks the keyboard status using the library function **KeyPressed**. If characters are waiting to be read, **Pro-**

cessKbd fetches the next keyboard character using the Turbo library function **ReadKey** which provides unfiltered and unechoed character input (it calls the BIOS keyboard routine). If the character is a normal key it is put into the transmit buffer and also written to the screen if local Echo is true.

Extended keys—the Alt, function, or cursor ("arrow") keys—are handled separately. **ReadKey** returns NUL (zero) if the character is an extended key; a second call to **ReadKey** then returns the scan code.

Because control characters are legitimate characters, Alt keys are used for program control as is usual in PC communication programs. The sample program implements only Alt-E to toggle the local echo and Alt-Q to quit.

The cursor keys send the appropriate ANSI cursor movement escape sequence and also move the screen cursor if local Echo is on. Function keys F1 through F4 send the DEC codes for the four VT100 function keys PF1 through PF4. The DEC "alternate keypad" codes are not implemented in the sample program.

Display

Only the most basic screen manipulations are provided in TinyTerm in its present form, using simple calls to the Turbo Pascal Crt Unit library. These can readily be expanded: for example, to implement "extents" (scrolling windows). Because the VT100 is a twenty-four-line terminal, line 1 or line 25 on the PC screen could be used for a status bar.

Escape sequences

ANSI escape sequences begin with a command sequence introducer (CSI) that can take several forms. The simplest is just the ESC character. The compound CSI has [as the second character for standard ANSI codes, or ASCII characters from : to ? for private nonstandard codes specified by each vendor. The sequence then has zero or more parameters, always numeric, which might be prefixed with a private code character. The last character is a terminal character from the ASCII characters @ to ~ that specifies the kind of operation to be performed.

The ANSI escape sequence parser is similar to a recursive descent parser. The path of calls to subroutines has something of the form of a tree, branching as required to handle the various forms of the escape sequence, as execution proceeds through the sequence. Each section of the escape sequence, activates a processing procedure until a terminal character is encountered, after which the corresponding operation is carried out. The numeric parameters, or arguments, are read off and converted to binary integers by the procedure **Arg** and stored in an array. **Arg** calls itself recursively until all arguments have been read. The function **NextArg** retrieves values in order from the parameter array. Because ANSI sequences permit the omission of parameters, **NextArg** is called with an appropriate default value that it returns if there is no value in the parameter array.

Extensions and additions

TinyTerm is meant primarily as working sample code to illustrate I/O techniques, not as an example of how to write a terminal emulator. Nevertheless, it is a sound framework on which to build if you would like to create a more capable one. An important addition would be to add code for flow control; the listing suggests where such routines would be added. Another addition needed to make a useful VT100-like emulator is code to implement the cursor and numeric keypad keys, which can be done in much the same way as the function key code already in TinyTerm. The **SetMode** and **ResetMode** routines should also be expanded. TinyTerm's main command execution subroutine implements only the most basic ANSI commands, but shows the lines along which additional commands could be added.

Summary

This chapter looked at serial I/O from soup to nuts, so to speak: from the PC serial port hardware and how to program it to routines for fast interrupt-driven serial I/O. You saw how a group of low-level service routines could be put together in a module, the serial toolkit, that application programs can use to handle the nuts and bolts of serial I/O in a systematic way.

Much of the material in this chapter is of very general application. The ways in which I/O functions can be incorporated into programs applies to all sorts of I/O, not just to the serial port. Polled and interrupt-driven methods were compared, and how both can be used together was discussed. Another important and general discussion covered interrupts, especially hardware interrupts, and how to write and use ISRs. These principles will be used in succeeding chapters.

5 ❖ The PC printer port

This chapter discusses how to use the standard parallel port for input and output. The PC supports up to three parallel ports, designated LPT1 through LPT3. The PC parallel port is designed primarily for output to printers using the Centronics interface. Most of us think of it only in that way, as the printer port, but it can be used for input too. It has several input lines, normally used to report printer status, and its printer control lines can be set for use as inputs. With proper programming the parallel port can provide quite respectable general I/O performance. Some file transfer utilities use the parallel port in this way.

Using the printer adapter as an input port is complicated somewhat by a lack of uniformity in the hardware. (You probably know only too well that things in the wonderful land of computers are seldom as straightforward as they could be.) The design of the original PC printer port has several flaws, the most serious of which is that it does not implement its hardware interrupt correctly. Later versions of the printer port often found in AT and 386 machines fix some of the problems. Some designs, such as those in many models of the IBM PS/2, even allow the port to be configured for true bidirectional operation. In discussing the parallel port it is necessary to take these variations into account. This applies to newer machines also, because inexpensive parallel port adapter cards seem to follow original IBM design fairly closely.

When reading this chapter you might begin to wonder if dealing with the quirks of the hardware on the one hand, and working around the fact that the port is one line shy of the eighteen lines needed for full simultaneous eight-bit strobed I/O, is really worth it. You might feel, not unjustifiably, that there is a little too much tinkering going on. Is it worth it? The answer, I believe, is yes. There are some applications for which the only workable options are either using the parallel port or installing a special I/O interface card in the computer. The latter might be the best approach in some cases. If it is desirable or necessary to be able to use different computers in the system, however, either the I/O card must be swapped or each computer must have such an I/O card—which might not be possible with a laptop computer or with an XT if the card requires a sixteen-bit system bus. In such cases, using the parallel port seems to be the only reasonable course.

The Centronics printer interface

Because the parallel port is designed specifically for driving printers, let's begin with a look at the Centronics printer interface. It consists of eight data lines plus a data strobe, an acknowledge line, and three control and four status lines. The lines and their customary names are as follows:

1. Output lines (computer to printer):
   - **D0–D7:** Data lines, positive logic (0 = low, 1 = high).
   - ***STB:** Data strobe pulse, negative-going; setup, strobe width, and hold times to be 500 ns or more (1 μs is typical). It loads the data on the data lines into the printer.
2. Handshake line (printer to computer):
   - ***ACK:** Acknowledge pulse, negative-going; nominal width 5 μs. Issued by the printer when data has been received and latched.
3. Status lines (printer to computer):
   - **BSY**: High when the printer is not ready to receive for any reason, such as during data latching, when the printer buffer is full, when the printer is off-line or out of paper, and so on.
   - **PE:** (paper end) High if the printer is out of paper.
   - **SEL:** High if the printer is selected (on-line).
4. Control lines:
   - ***ERROR:** (printer to computer) Low if there is an error or fault condition in the printer.

- ***AUTOLF** or ***AUTOFEED:** (computer to printer) Low causes automatic line feed after carriage return charac - ter (no line feed character needed).
- ***INIT:** (computer to printer) Low resets printer and clears the printer buffer.
- ***SELINPUT** or ***SELIN:** (computer to printer) Low to enable printing.

The standard Centronics interface uses Amphenol 57 series thirty-six-pin connectors, female on printers and male on cables (57-30360 or equivalent). IBM chose instead to use a twenty-five-pin DB-25 connector for the parallel ports of the PC. Figure 5-1 shows the pin assignments used on both connectors. Electrically the lines are standard TTL-level: low is less than 0.8 V and high is greater than 2.0 V. There is usually no effort to provide reasonable termination or impedance matching (a topic taken up in chapter 7), so the maximum cable length for satisfactory performance is

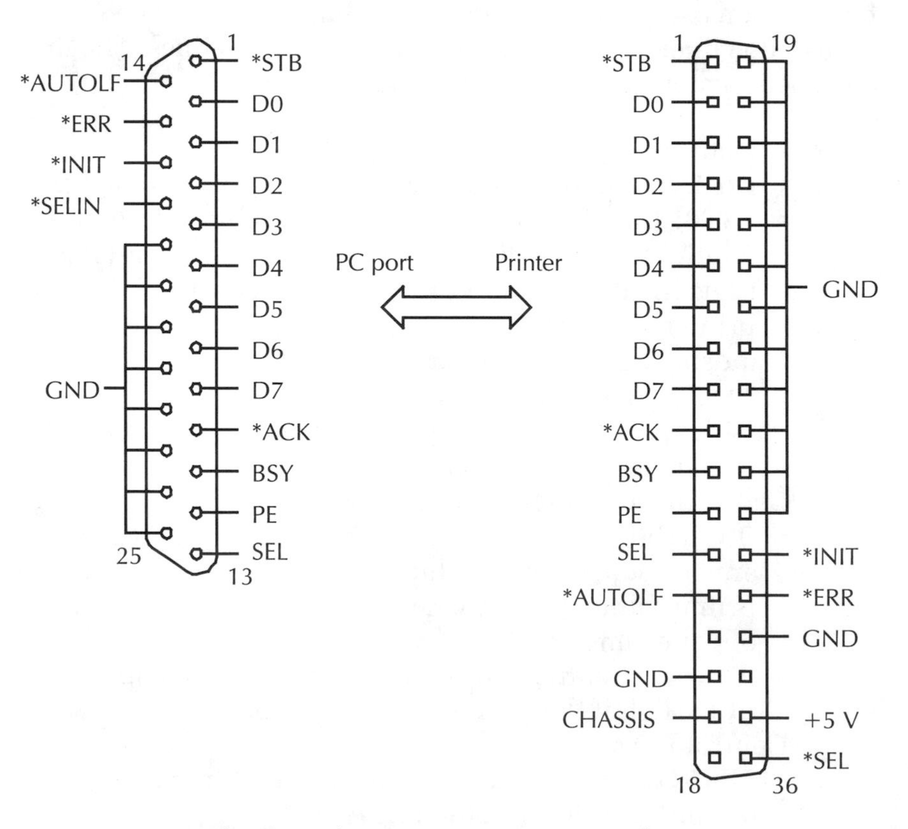

Fig. 5-1 The Centronics printer interface.

usually on the order of 10 to 12 feet. The fast rise and fall times of the signals can generate a significant amount of EMI (RF noise), so shielded cables should be used. If for some reason ribbon cable must be used, it should be the twisted-pair variety, not the plain flat type. Commercial ready-made printer cables work well, are convenient, and are recommended.

The parallel port adapter

A block diagram of the original IBM parallel port adapter is shown in Fig. 5-2. Later versions differ in a number of respects, but for the most part they are functionally equivalent. In keeping with its design as a printer driver, the adapter hardware is organized into three functional blocks: data, status, and control. There is also a section devoted to the interface with the PC system bus containing I/O address decoding and read/write control circuits.

The data section consists primarily of an eight-bit latched bus driver (LS374) that holds the current output byte. The eight output lines (D0–D7) are also connected to a three-state buffer (LS244) that allows the output lines to be read by software. The LS374 latched driver is a three-state device, permanently enabled. This means that the only thing that can be read from the data section is the byte currently in the output latch. Had the enable pin of the LS374 latch been subject to control by software—there are enough unused control bits to accommodate this—the port could have been designed so that the

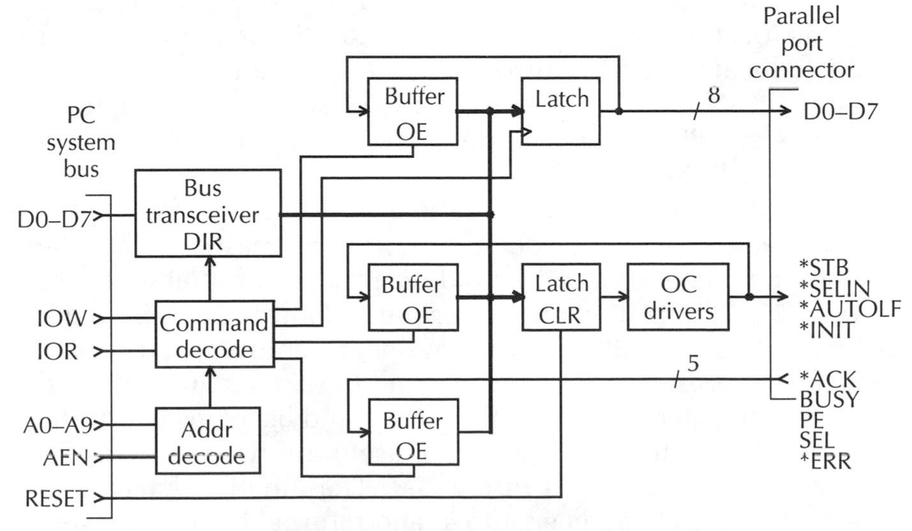

Fig. 5-2 Block diagram of the original PC parallel port.

driver latch could be disabled to allow the LS244 buffer to respond to the eight data lines as true inputs. It is possible to modify a printer adapter card to do this, but it involves cutting circuit board traces and adding jumpers. The details are left as an exercise for interested readers because this book abstains from any tinkering with the inside of the computer.

The status section consists of a set of buffers that report the current levels of the status lines (BSY, PE, SEL, and *ERROR) when enabled. These lines are input only. The status buffer also reads the *ACK line, which is connected through a three-state buffer to a system bus (IRQ) line, usually IRQ7.

The control section contains four drivers that drive the printer control lines (*SELINPUT, *INIT, *AUTOLF) and the *STB line. Input buffers are connected in parallel with each driver so the state of the lines can be read. The drivers are open-collector types, which means that they are effectively out of the circuit when they are at high output level. When thus set, the control lines become true inputs and can be used as such. Some care has to be exercised in using these lines and the status lines for input because some of the lines are inverted (a high line level produces a zero bit and vice versa) and others are not. This is considered in more detail a little later.

The organization of the parallel port adapter is reflected in the arrangement of the interface to the PC system bus. The parallel port appears as three I/O registers. There does not seem to be a standard nomenclature for these registers, so the following conventions are used in this book. The printer data register (PDR) is located at the base address. The printer status register (PSR) is located at I/O address (base + 1). The printer control register (PCR) is at (base + 2). The PDR is a straight eight-bit I/O register. The other registers are bit-significant and are listed in Table 5-1.

Unlike the serial ports, there is a certain ambiguity in the base I/O addresses of the parallel ports. The reason is that the monochrome video adapter includes a printer port that is at base address 3BCh, and the straight printer adapters have base addresses of 378h and 278h. During system initialization, the BIOS tests for the presence of hardware at addresses 3BCh, 378h, and 278h in that order. The first valid port found becomes LPT1, the second LPT2, and so on. A couple of useful routines to check for parallel ports is presented in Fig. 5-3. The function **ParBase** is called with an LPT number, 1 to 4, and returns the base address

Table 5-1 Parallel port register bit assignments.

Bit	Function
Data register (PDR), at base address	
	Write: sends data byte to printer
	Read: returns the byte last written
0–7	Data
Status register (PSR), address (base + 1)	
	Read only
Bit	**Function**
7	BSY: 0 = busy, 1 = not busy
6	*ACK
5	PE: 0 = okay, 1 = out of paper
4	SEL: 0 = printer off-line, 1 = on-line
3	*ERROR: 0 = error, 1 = okay
2	Not used (on some versions: IRQ has occurred)
0–1	Not used
Control register (PCR), address (base + 2)	
	Write: sets control lines and hardware interrupt
	Read: returns current status of the lines
7–5	Not used
4	0 = disable interrupt, 1 = enable interrupt
3	*SELIN: 0 = deselect printer, 1 = select
2	*INIT: 0 = initialize printer, 1 = normal operation
1	*AUTOLF: 0 = normal (requires CRLF), 1 = auto LF (requires CR only)
0	*STB: 0 = strobe off (STB high), 1 = assert strobe (STB low)

from the BIOS table. An address of 0 means that there is no port for that LPT number. (A small utility program that checks for all parallel and serial ports is given in chapter 7.) The function **TestParHdwe** tests for the presence of actual parallel port hardware at the base address passed to it. Because the data written to the PDR register can be read back, a byte is written and then the PDR is read. To be on the safe side, this is done twice with a zero byte and an all-ones byte.

The BIOS table method might fail in certain special cases. If the computer is being used as a network file or printer server, for

Fig. 5-3 Routines to check for the presence of a parallel port.

```
{ Example of how to get parallel port address from BIOS table. }

function ParBase( N : byte ) : word;                    { 1 = LPT1, etc. }
var
   Table : array [0..3] of word absolute $40:0008;
begin
   N := (N-1) and 3;                                  { adjust N and clamp }
   ParBase := Table[N];                                       { get value }
end;


{ Routine to verify presence of parallel port by hardware test. }

function TestParHdwe( Base : word ) : boolean;
var
   Save,T1,T2 : byte;
begin
   Save := Port[Base];                           { get and save current value }
   Port[Base] := $FF;                              { write first test value }
   T1 := Port[Base] ;                                               { read }
   Port[Base] := 0;                                          { second test }
   T2 := Port[Base];
   TestParHdwe := (T1 = $FF) and (T2 = 0);                        { result }
   Port[Base] := Save;                            { restore current value }
end;
```

example, the network shell might set some or all of these port base addresses to zero even though the hardware exists. The idea is that these ports are used or reserved for network printers and therefore should be marked as nonexistent so that applications will not try to use them.

The parallel port provides a hardware interrupt, activated by the *ACK line going low. Hardware interrupt IRQ7 is standard for the LPT1 device, but beyond that the situation is somewhat confused. Many AT and 386 machines use IRQ5 for LPT2. In XT machines, however, IRQ5 is used for the hard disk, so in XT computers IRQ2 is often used for LPT2. Things are sometimes further complicated by the fact that IRQ2 is commonly used by LAN cards in both XTs and ATs, and so might not be available if the computer is on a network.

Adding to the problem with parallel port interrupts is the fact that the design of the original parallel port adapters is flawed. Recall from the discussion of hardware interrupts in chapter 2 that asserting an interrupt request (IRQ) line raises the INTR pin of the CPU only after passing through the interrupt controller (PIC). The PIC requires the IRQ to remain asserted until it is acknowledged by the CPU. This typically takes about 10 to 15 μs in a typical AT, and 80 to 90 μs or more in a 4.77-MHz PC. The interrupt latency might be even greater if other higher priority interrupts are pending or software has temporarily suspended interrupts during a critical section of code. The mistake in the original design is that the incoming interrupt signal, which is the Centronics *ACK line, is simply passed on to the computer's IRQ line. If the *ACK signal is shorter than the interrupt latency, the process will misfire. When the port is being used with a printer, the standard *ACK pulse is about 5 μs and the printer interrupt will not work correctly. It is for this reason that printers are hardly ever interrupt-driven; polling is used to check the printer BUSY status instead.

The interrupt problem could have been avoided if the *ACK line had been latched (used to set a flip-flop). The problem doesn't exist in the serial adapter because the 8250 UART has built-in interrupt latches. Later versions of the parallel port, particularly in AT, 386, and 486 machines, often do latch the *ACK input. Some allow selection of whether to latch on the leading (negative-going) or the trailing (positive-going) edge of the pulse. Unfortunately there is no simple way to tell whether a machine has such a port, short of testing it or reading the computer's documentation (if it can be found). In the interrupt-driven input routines described later in this chapter you will be reduced either to using fairly long strobes, which wastes time in the case of an AT or 386, or to the inelegant procedure of trying a short pulse and if that doesn't work trying longer pulses until reliable operation results.

Parallel port I/O

The parallel port provides both input and output functions. The data output is through the data register (PDR) which maps the D0–D7 bits of the data byte directly to output pins 2 through 9. The output strobe, *STB, is controlled by bit 0 of the control register (PCR). The *STB pulse must be created by software; the *STB output line remains in the same state until it is changed. The other output lines are controlled by other bits in the PCR as shown in Table 5-1.

The internal layout of the various inputs is shown in Fig. 5-4. The input lines are read by reading the status register (PSR) or the PCR as the case may be. Only the BSY, PE, SEL, *ERROR, and *ACK lines are purely inputs. The remaining lines are combination inputs and outputs, driven by open-collector logic devices. If the outputs of these drivers are set to high, they are effectively out of the circuit. On each line there is a pull-up resistor (usually 4.7 kΩ) connecting the line to +5 V. When the driver is in the high state, the resistor pulls the line up to high level. So long as the driver output remains at the high state, the *STB, *AUTOLF, *SELINPUT, and *INIT lines can be used as inputs. You will notice that the overall logical polarity is not the same in all of the inputs in both groups. BSY, *STB, *AUTOLF, and *SELINPUT invert; the others do not.

In order to use any of the lines connected to the PCR, the output drivers must all be preset for high output before the lines can be used for input. The drivers are controlled by the lower four bits in a byte written to the PCR. Bit 5 enables the hardware interrupt if it is set to 1, or disables it if cleared to 0. To preset the drivers to use the control lines as input, write a byte to the PCR (at address base + 2) as follows:

- For disabled interrupt: write 04
- For enabled interrupt: write 14 hex (20 decimal)

Using outputs as inputs applies only to the lines just mentioned. Do not try to use the data lines (pins 2 through 9) as inputs. (The standard printer adapter does use a three-state driver

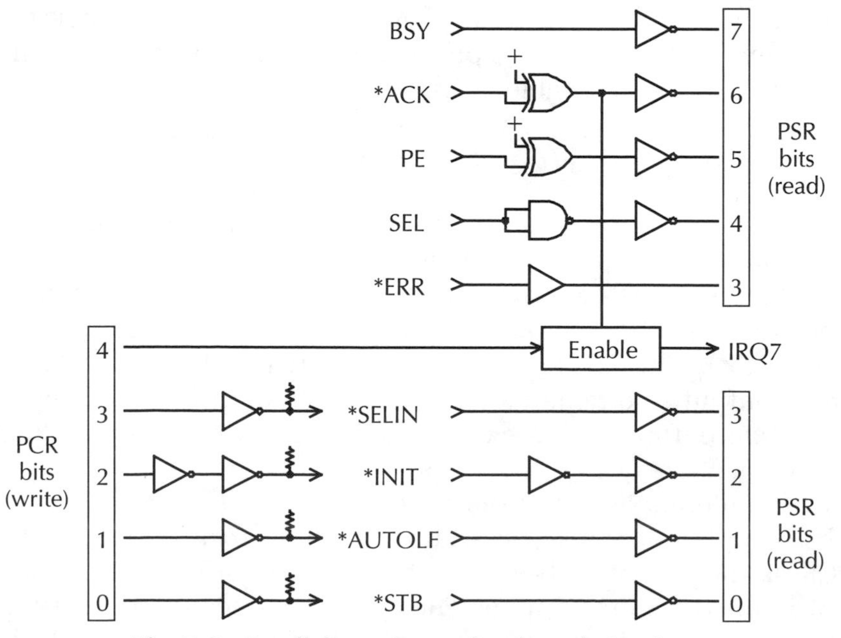

Fig. 5-4 *Parallel port input signals and circuits.*

for data output, but there is no way short of cutting traces on the adapter circuit board to set it to the OFF state to allow D0–D7 to be used as inputs.)

Please notice also that the PCR lines cannot safely be used as inputs if there is any chance that the BIOS or DOS printer or parallel port routines will be used. They use the *STB line for the data output strobe and set the other lines as appropriate for printers. Unfortunately, the PSR lines cannot be used for general input either because the PSR is read each time a byte is sent in order to check for printer busy, out of paper, or other errors. The BIOS routine would interpret any data on those lines as printer status information. You will see presently how to create strobed data output without going through the BIOS or DOS.

Using the parallel port for output

The job for which the parallel port was designed is output, specifically to printers using the Centronics interface. The software drivers provided in the BIOS and DOS assume, not unreasonably, that the port will be used for that purpose. Nevertheless the port can indeed be operated as a general-purpose parallel output port. Let's consider all of these uses.

Using the BIOS driver

The BIOS parallel port services run the port specifically as a printer interface, and they are very convenient (indeed, the method of choice) for that purpose. Although this is a book about data acquisition, printing is certainly not unimportant in such systems. A brief consideration of the parallel port as a printer interface is therefore not out of place.

The BIOS routines are also quite useful for output to external devices containing a Centronics-like interface that emulates a printer, a topic discussed further in chapter 6. The BIOS driver cannot be used at all if the port is being used for data input. How output is done in that case is discussed later in this chapter.

In most cases, the best way to use the parallel port for printing is through DOS as a file, using one of the predeclared printer device names (PRN, LPT1, LPT2, LPT3). A nice feature of this method is that printer output can be directed to a disk file merely by substituting a normal disk file name in place of the device name. Ordinary text file output routines can be used for printer output, but there is a much more efficient way, particularly in the case of data that is already in block form. The better way is to use an output function that directly calls the DOS stream output function 40h, write to file or device. Many high-level languages provide such output routines, for example **BlockWrite** in Turbo Pascal or **_write** in Turbo C.

As mentioned in chapter 2, printing binary data such as bitmapped graphics sometimes doesn't work properly if the system traps certain control characters, so the DOS IOCTL function may have to be used to set "raw" output mode.

When parallel output is done through DOS, the DOS critical error handler (interrupt 24h) will report errors, usually either error 09, out of paper, or 0A, write fault (if the printer is off line, for example). The extended error call in DOS 3.0 and later, DOS function 59h, maps these errors to 1Ch and 1Dh respectively. The Turbo Pascal *IOresult* variable maps them to 159 and 160 (decimal) respectively.

A less flexible but in some ways superior approach is to use the BIOS printer services directly. The BIOS printer output routine will do only one character at a time, but there is no funny business with the trapping of control codes. The BIOS printer functions are called through interrupt 17h with the DX register set for the port to use, starting at 0 for LPT1. The BIOS provides three functions, selected by the value placed in the AH register: write byte (AH = 0), initialize printer (AH = 1), and report printer status (AH = 2). For write byte, the character to be written is put in the AL register. All functions return with the current printer status in register AH.

Which method is better? The correct answer, as so often with computers, is "It depends." I myself use both. The line drawings in this book were created with a commercial CAD program used as a "front end" for a program I developed for doing vector graphics on a Hewlett-Packard LaserJet III printer. Because HPGL graphics commands consist almost entirely of printable ASCII characters, printing through DOS works just fine and seems the better choice. On the other hand, for printing bitmapped graphics and for downloading printer fonts I always use the BIOS routines directly.

The BIOS printer status function is quite useful even when output is being done through DOS. A handy routine is shown in Fig. 5-5. The function **PrnStatus** is called with the LPT number (1 = LPT1, and so on). The bits in the BIOS status byte are manipulated a little so that the function returns 0 if the printer is ready to go. Nonzero values indicate one or more errors as listed in Fig. 5-5. This function is especially useful for making sure that the printer is on-line before starting printing, and for checking that the printer hasn't run out of paper before beginning printing to a new page.

Direct register output

Parallel port output can be done by direct I/O to the port registers instead of through the BIOS or DOS. In fact, when the parallel port is being used for input it must be done directly. Because this involves going directly to the hardware, it is a little more complicated than going through the DOS or BIOS. There are three cases to consider depending on how the parallel port is being used.

The simplest case is when the output lines D0–D7 are being used as static outputs. Writing a byte to the PDR at the base I/O address of the port sets the output lines, and those output levels remain where they are until another byte is written to the PDR.

More work needs to be done when the output from the port requires a data strobe or "data ready" output also. Let's consider first the case in which the *STB line is being used in the usual way as an output strobe line. The output routine must write the output data to the PDR and must then manipulate the *STB line through the PCR to create a strobe pulse. At the same time it must not change any of the other lines in the PCR, leaving them free for use as inputs. The procedure **Strobe8** in Fig. 5-6 shows how

Fig. 5-5 *A routine for checking printer status.*

```
{ Routine to check printer status. }

USES Dos;

{ PrnStatus returns 0 if printer is ready, or exceptions as follows: }
   { bit 0: 1 = timeout error }
   { bit 3: 1 = I/O error     }
   { bit 4: 1 = off line      }
   { bit 5: 1 = out of paper  }
   { bit 7: 1 = busy          }

function PrnStatus( P : byte ) : word;          { P : 1=LPT1, 2=LPT2, 3=LPT3 }
var
   Regs : Registers;
   B : byte;
begin
   Regs.AX := $0200;
   Regs.DX := (P and 3)-1;            { AH = 2 (get status); AL = device, 0..2 }
   Intr( $17,Regs );
   PrnStatus := (Regs.AH and $B9) xor $90; { mask unused bits; invert BSY,PE }
end;
```

it can be done. Note that the port addresses are held in global variables, which must be set before calling the routine. The listing shows the most common case, LPT1 at base address 378h. **Strobe8** is functionally equivalent to the BIOS print byte function except that it does not check any input status lines and thereby leaves them free for use as input lines.

The duration of the strobe pulse generated by **Strobe8** depends on many factors, among them the CPU type and clock rate and the I/O bus rate. On most machines the routine will generate a pulse a few microseconds wide, which is about right for most applications. It can be checked with an oscilloscope if the exact pulse width is important. If it is too short, some "do-nothing" code (like incrementing a scratch variable) can be inserted to keep *STB at low level longer. The listing in Fig. 5-6 includes a little test program that writes a text string to a standard printer. A loop sends out the characters, first checking the printer BUSY status and sending the character using **Strobe8** if the printer is ready. If the printer prints the string, the strobe pulse can be assumed to be of reasonable duration. Don't forget that many whole-page printers such as the HP LaserJet require a form feed (^L; 12 decimal, 0C hex) after the text in order to print the page.

The third situation arises when the *STB line is being used as

Fig. 5-6 *Routines for strobed output through the parallel port.*

```
{ Routines for strobed output through the parallel ports. }

{ The following global variables must be set for the port to be used.  }
{    Example: for LPT1, ParBase := $378; ParSR := $379; ParCR := $37A; }

VAR
   ParBase, ParSR, ParCR : word;


procedure Strobe8( D : byte );
var
   CR : byte;
begin
   Port[ParBase] := D;                          { write data to Data Register }
   CR := Port[ParCR];                              { get and save Control Reg }
   Port[ParCR] := CR or 1;                                     { set STB low }
   Port[ParCR] := CR;                                         { set STB high }
end;


procedure Strobe7( D : byte );
begin
   Port[ParBase] := D or $80;                     { write data with D7 high }
   Port[ParBase] := D and $7F;                      { set D7 low to strobe }
   Port[ParBase] := D or $80;                           { restore D7 high }
end;
```

an input. This case is looked at in more detail in the next section. In this case, one of the data lines must be borrowed and used as a data output strobe or a data-ready line. The procedure **Strobe7** in Fig. 5-6 handles this case. Data line D7, the most significant bit, is used for the strobe. Note that only seven-bit data can be transmitted. This might not be a significant limitation if the computer output is being used for sending standard seven-bit ASCII characters or to control an external device.

Both of the routines in Fig. 5-6 generate data strobe pulses. If your application uses a static data-ready status line approach you can easily modify the routines so that they write the data and then leave the data-ready line in the asserted state. A routine to reset the data-ready line to the not-asserted state would need to be written also.

Using the parallel port for input

The layout and general characteristics of the input lines of the parallel port have already been discussed. Because the adapter was not designed with ordinary data input in mind, the inputs have to be manipulated in software to some extent to accommodate general data input. Aside from this small additional complication, however, the standard parallel port works very well for input. Several different arrangements of the inputs are possible. Which arrangement is best depends on the requirements of each particular application. Although the number of different configurations might seem slightly bewildering on a first reading, they do fall into just a few categories.

When any of the parallel port status or control lines are used for input it is important to take into account the fact that some lines have normal logic sense—the bit is 0 for a low input level and 1 for a high level—whereas other lines are inverted. Table 5-2 lists the logical sense of the status and control lines. To obtain the correct result the bits of any inverted lines must be "flipped" or reinverted. The XOR operator is an efficient way to do this. XORing an inverted bit with a 1 bit flips it.

Using the status lines for input

The status lines are input only, and they can be used for either polled or interrupt-driven input. In interrupt-driven operation, *ACK is used as a strobe to activate the IRQ line and the remaining four lines are the data inputs. This is well suited to four-bit

Table 5-2 Logic sense of parallel port inputs.

Bit	Name	Sense
Status register (PSR), address (base + 1)		
7	BSY	Inverted
6	*ACK	Normal
5	PE	Normal
4	SEL	Normal
3	*ERROR	Normal
Control register (PCR), address (base + 2)		
3	*SELIN	Inverted
2	*INIT	Normal
1	*AUTOLF	Inverted
0	*STB	Inverted

BCD applications. In polled operation, the *ACK line can be used either as a status line or as another data line, giving five data input lines.

Status-line data is obtained by reading the PSR at address (base + 1). The bits in the status byte are not located where they would be in a numerical byte, however. A look at the bit assignments in Table 5-1 shows that five-data-line operation could fix this by means of a three-bit right shift to right-justify the bits. Table 5-2 shows that bit 7, the BSY line, is inverted; it must be flipped to restore the correct logical sense.

A little more work has to be done when the *ACK line is not used for data input but as a signal line (data-ready status or interrupt strobe) because this leaves a "hole" in the PSR byte at bit 6. A good way to handle this is as follows. First, the *ACK bit—bit 6—is replaced with bit 7 (BSY). Because bit 7 is inverted it must be flipped. Second, the result is shifted right to right-justify the bits. A fairly concise routine to do all this is illustrated in the **PSRinput** function in Fig. 5-7. The code also masks off all unused bits to head off any unpleasant surprises. The global variable *ParSR* is used to hold the address of the PSR (base + 1).

The **PSRinput** function can be used in conjunction with a routine such as **ACKlow,** also in Fig. 5-7, to check the status of the *ACK bit for polled operation. If the *ACK line is used as an interrupt trigger (as discussed later in this chapter) the **PSRinput** function can be expanded into an ISR to provide four-bit interrupt-driven input.

Fig. 5-7 Routine to read the status register lines.

```
{ Routines for parallel port status byte input. }

{ The following global variable must be set for the port to be used.  }
{    ParSR is ParBase+1. Example: for LPT1, ParSR := $379;             }

VAR
   ParSR : word;


function PSRinput : byte;
var
   B : byte;
begin
   B := Port[ParSR];
   if (B and $80 <> 0) then PSRinput := (B and $38) shr 3 else
      PSRinput := ((B or $40) and $78) shr 3;
end;

function ACKlow : boolean;               { returns TRUE if -ACK line is LOW }
begin
   ACKlow := (Port[ParSR] and $40 = 0)
end;
```

Using the control lines for input

The control lines can be used for four-bit input by reading from the PCR at address (base + 2). The bits are already in a satisfactory position so no shifting needs to be done, but as Table 5-2 indicates only bit 2 (*INIT) has normal logic sense. The other bits need to be flipped, so the byte should be XORed with binary 1011 (0B hex). The byte from the PCR should also be ANDed with 0F hex to mask off all the unused bits. Before using these lines for input the open-collector output drivers connected in parallel with the inputs must be set high by writing the value 04 to the PCR.

Using the control and status lines for eight-bit input

The two four-bit "nibbles" from the PSR and PCR can be combined to provide full eight-bit input. Together with the standard output this provides for bidirectional eight-bit I/O operation, with certain qualifications.

The reason for the qualifications is that the parallel port provides seventeen lines altogether. This accommodates two eight-bit data paths but it is one line shy of two separate data signal paths. (Recall the distinction made in chapter 3 between a "blind" parallel link and one with a "signal," a status or strobe, line.) If eight-bit operation is required for both input and output, then either the output or the input signal line must be sacrificed. If both an output signal and an input signal are needed, then either the input or the output link will have to be reduced to seven data bits.

A good way to classify the possibilities is in terms of whether the optimum path—the primary or most critical one in terms of performance—should be output from the PC, or input to the PC, or both. The path with the higher priority gets to use both full eight-bit data and a signal line (status or strobe). The combinations are set out in Table 5-3.

Table 5-3 Parallel interconnection configurations.

Primarily input to PC:
8-bit input with signal; and either
7-bit output with signal *or*
8-bit blind output
Primarily output from PC:
8-bit output with signal; and either
7-bit input with signal, *or*
8-bit blind input
Equal input and output priority:
8-bit input,
8-bit output, and
1 extra line

Eight-bit I/O primarily for input

Let's consider first the input case in which the primary data flow is into the computer from an external device (or another PC). The *ACK line is used for the data signal, which can be either a status signal in polled operation or a strobe for interrupt-driven input.

The connections to the parallel port for input are shown in Fig. 5-8. The arrangement shown requires the least amount of bit shuffling. Standard printer cables are handy for connecting external devices to the PC parallel port, and for your convenience the pin connections to a 57-30360 (or Centronics) connector are shown in Fig. 5-9. The external device must provide suitable interface circuits such as those shown in chapter 6. If the source of data is another PC they can be cross-connected as indicated in Fig. 5-10.

Don't forget that the open-collector output drivers connected in parallel with the PCR inputs must be set high by writing the value 04 to the PCR, I/O address (base + 2), before using the control lines for input.

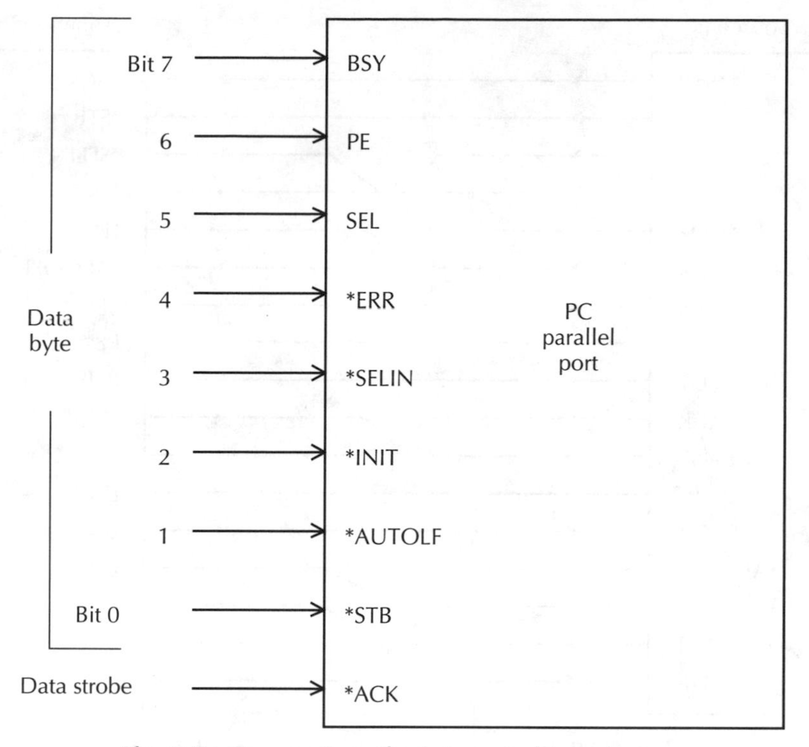

Fig. 5-8 *Connections for interrupt-driven input.*

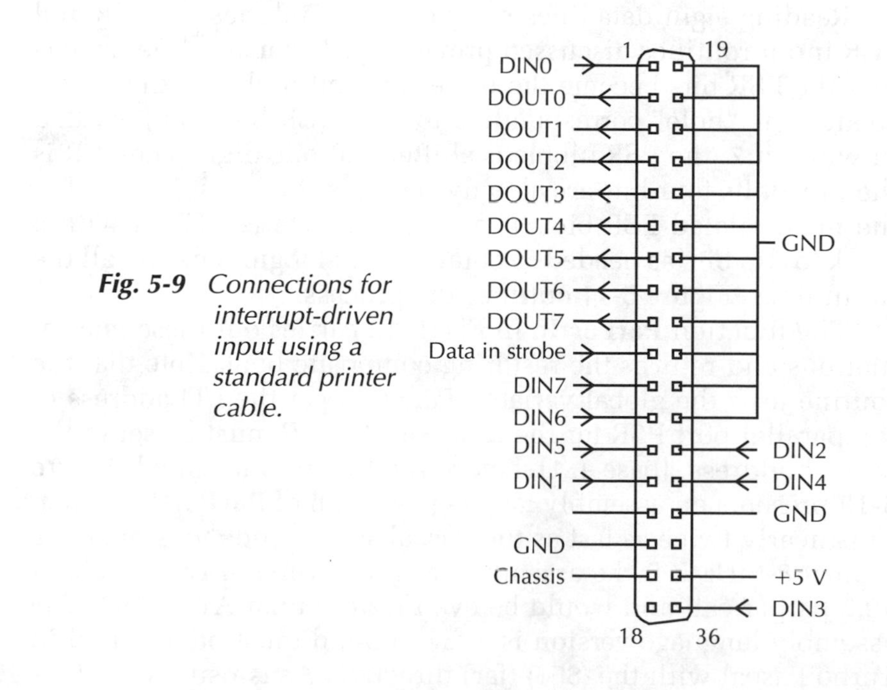

Fig. 5-9 *Connections for interrupt-driven input using a standard printer cable.*

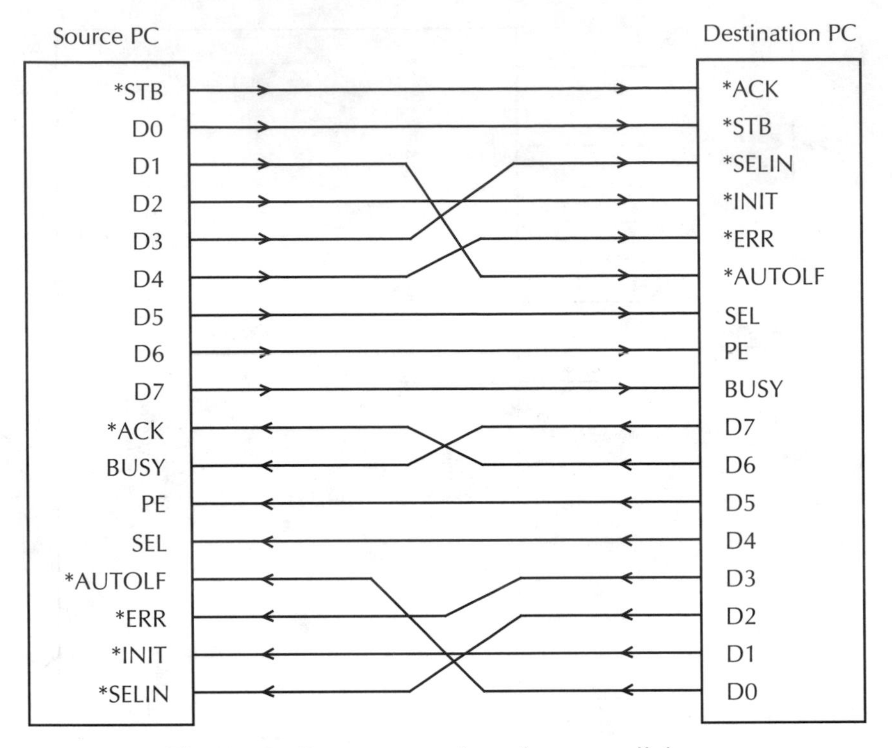

Fig. 5-10 *Cross-connection of two parallel ports.*

Reading eight data lines more or less combines the PSR and PCR input routines discussed previously. The main difference is that the PSR bits become the upper four-bit nibble of the result, so after the "hole" corresponding to the *ACK bit (bit 6) is filled in with bit 7, the PSR nibble is shifted left one digit. The PCR is then read. Its four bits are already right justified so it is joined to the manipulated PSR bits. The resulting byte is XORed with a mask to flip bits as needed to obtain normal logic sense on all the input lines. Figure 5-11 outlines the process.

The function **ParPortIn** in Fig. 5-12 carries out these manipulations and returns the resulting composite byte. Note that the routine uses the global variable *ParSR* to get the I/O address of the parallel port PSR for the first read. *ParSR* must be set to the correct address, (base + 1), before the function is called. Figure 5-13 presents an assembly language version of **ParPortIn**, which runs nearly twice as fast as the Pascal source code version. On a typical 8-MHz XT, the assembly language version runs at about 49.5 μs per call and would be even faster on an AT or 386. The assembly language version is a far call and must be declared in Turbo Pascal with the {$F+} (far) directive. As is usual with Pas-

```
1. Get PSR byte
  bits    7      6      5     4      3     2   1   0
        | BSY | *ACK | PE | SEL | *ERR | ? | ? | ? |

Put a copy of PSR bit 7 where bit 6 (*ACK) was
        | BSY |  BSY | PE | SEL | *ERR | ? | ? | ? |

Shift left 1
        | BSY | PE | SEL | *ERR | ? | ? | ? | 0 |

AND with F0h
        | BSY | PE | SEL | *ERR | 0 | 0 | 0 | 0 |


2. Get PCR byte
        | ? | ? | ? | ? | *SELIN | *INIT | *AUTOLF | *STB |

AND with 0Fh
        | 0 | 0 | 0 | 0 | *SELIN | *INIT | *AUTOLF | *STB |

OR with modified PSR from step 1
        | BSY | PE | SEL | *ERR | *SELIN | *INIT | *AUTOLF | *STB |

Flip bits (XOR with mask)
        | BSY | PE | SEL | *ERR | *SELIN | *INIT | *AUTOLF | *STB |
mask       1     0     0     0       1        0        1        1


3. Result is final output
     | /BSY | PE | SEL | *ERR | /*SELIN | *INIT | /*AUTOLF | /*STB |
```

Fig. 5-11 *Building an eight-bit byte from the PSR and PCR bits.*

cal, the function must be identified as an **external** routine. (If you are linking the assembler version with another language such as C, note that the function returns the data byte in the AL register.) You might notice that the bit manipulations are slightly different from the example given in the preceding paragraph; these small changes increase efficiency. The comments in the assembler version should clarify the differences.

ParPortIn is for polled operation. The function **ACKlow** in Fig. 5-7 can be used to monitor the state of the *ACK line. In many cases it will be desirable or necessary to use interrupt-driven input. Discussion of interrupt-driven input is deferred until later, where **ParPortIn** will form the core of an ISR.

When the parallel port is used for eight-bit input in this manner only eight lines are available for output from the port.

Fig. 5-12 Pascal code to build an eight-bit byte from the PSR and PCR bits.

```
{ Pascal routine for 8-bit input from the parallel port using the Status
{ and Control registers. *ACK can be used for interrupt strobe.

CONST
   PolarityMask : byte = $8B;

{ The following global variable must be set for the port to be used.  }
{    ParSR is ParBase+1. Example: for LPT1, ParSR := $379;            }

VAR
   ParSR : word;


function ParPortIn : byte;
var
   B : byte;
begin
   B := (Port[ParSR]);
   if B and $80 = 0 then B := (B and $38) shl 1
      else B := ((B or $40) shl 1) and $F0;
   ParPortIn := (B or (Port[ParSR+1] and $0F)) xor PolarityMask;
end;
```

Writing a byte to the PDR at the port base address provides eight-bit blind output. If an output signal (strobe or status) line is required, one of the data output lines must be used because the *STB line is used as one of the input lines. The procedure **Strobe7** in Fig. 5-6, for example, uses data line D7 for an output strobe, with lines D0–D6 providing seven-bit data output.

Eight-bit I/O primarily for output

When the primary data flow is output from the PC, the eight data output lines carry eight-bit data and the *STB line is used for the output data signal, either as a status line or for a data strobe. The appropriate connections are indicated in Fig. 5-14.

Transmitting data from the parallel port consists of writing the data byte to be sent into the PDR at the port base I/O address. The *STB line is then set, usually to low to indicate valid data by clearing bit 0 in the PCR at address (base + 2). The procedure **Strobe8** listed in Fig. 5-6 is handy for producing strobed output. The BIOS and DOS output routines cannot be used for output because they manipulate the PCR and interpret the PSR specifically for printers. Because the *STB line is being used as an output strobe, only eight lines remain available for input. They can be deployed for either eight-bit blind input or for seven-bit input with a data signal (status or strobe).

For eight-bit blind input, all input lines except *STB are read.

Fig. 5-13 *Assembler code to build an eight-bit byte from the PSR and PCR bits.*

```
;------------------------------------------------------------
;            8-bit parallel port input routine
;------------------------------------------------------------
; PAR8IN.ASM 2.0 ©1992 J H Johnson

DATA    SEGMENT BYTE PUBLIC

        EXTRN ParSR : word

DATA ENDS

CODE    SEGMENT BYTE PUBLIC

        assume cs:code,ds:data

        PUBLIC ParPortIn

        ;------------------------------------------------------------
        ; function ParPortIn : byte; { far }
        ;------------------------------------------------------------

; equates
        PolarityMask    equ 10001011b   ; mask to flip bits

ParPortIn       PROC FAR

        push    bp
        mov     bp,sp
; get input and shuffle bits
        mov     dx,[ParSR]              ; Base+1 address - first read
        in      al,dx                   ; get data in AL
; "copy" bit 7 (pin 11) to bit 6 (pin 10) and shift left
        and     al,0B8h                 ; force bit 6 to 0, mask bits 0..2
        shl     al,1                    ; shift, carry flag gets old bit 7
        jnc     P1                      ; old bit 7 was 0, ok to continue
        or      al,80h                  ; old bit 7 was 1, so set bit 7
P1:     mov     ah,al                   ; ah will be upper nibble of result
        inc     dx                      ; Base+2 address - Control Reg
        in      al,dx
        and     al,0Fh                  ; keep lower 4 bits only
        or      al,ah                   ; add in upper nibble
        xor     al,PolarityMask         ; flip bits for correct sense
; return result which is now in AL
        mov     sp,bp
        pop     bp
        ret
ParPortIn       ENDP

CODE    ENDS
END
```

The PSR supplies the upper five bits. The PCR, after it is shifted one digit to the right to eliminate the *STB bit, supplies the remaining three bits. The result is XORed with 1000 0101 (85 hex) to obtain the correct logical sense. A concise routine to do this is given in Fig. 5-15. The function **ReadPar8** uses the global variable *ParSR* to get the I/O address of the PSR, (base + 1).

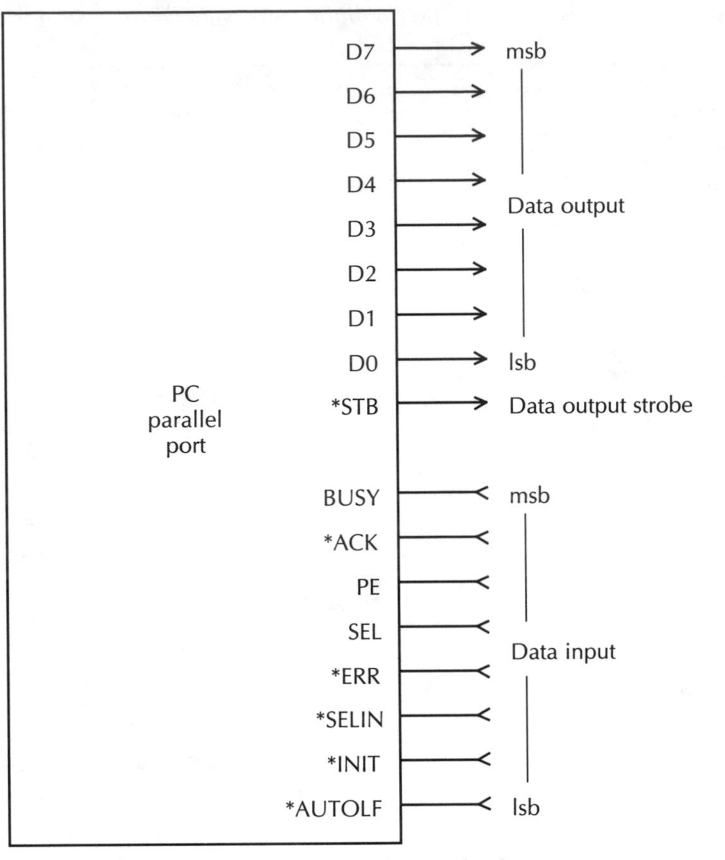

Fig. 5-14 *Connections for strobed output.*

Fig. 5-15 *Routine for eight-bit polled input.*

```
{ Routine for 8-bit input from the parallel port. *ACK is used for input }
{ data and *STB is unused and available for use as an output strobe.     }

{ The following global variable must be set for the port to be used.  }
{    ParSR is ParBase+1. Example: for LPT1, ParSR := $379;            }

VAR
   ParSR : word;


{ read BSY,*ACK,PE,SEL,*ERR, add *SELIN,*INIT,*AUTOLF; flip with 10000101b }

function ReadPar8 : byte;
begin
   ReadPar8 := ((Port[ParSR] and $F8) or
      ((Port[ParSR+1] and $0E) shr 1))
         xor $85;
end;
```

For seven-bit input with a data signal, the *ACK line is used for status or input strobe. The input can run in either polled or interrupt-driven mode. For polled operation, the function **ParPortIn** (Fig. 5-12 or 5-13) will work just fine with one minor adjustment. The *STB bit is bit 0, so simply shifting the byte returned by one digit eliminates bit 0 and moves the other bits to the correct place. The simplest way to do this is to operate directly on the value returned, like this:

```
Result := ParPortIn shr 1;
```

The same technique also works for bytes fetched from the input buffer in the interrupt-driven input routines described later. The ISR requires no modification.

Don't forget that the open-collector output drivers connected in parallel with the PCR inputs must be set high by writing the value 04 to the PCR, I/O address (base + 2), before using the control lines for input.

A special case exists when the interface in the external device has been designed specifically to emulate the Centronics interface of a standard printer. Circuit details are given in chapter 6. In this case, it is permissible, and usually desirable, to use the DOS or BIOS printer routines for output from the computer. The kinds of input to the computer from the external device are of course quite limited. The data flow is controlled by handshaking between the output strobe from the computer and the activation of the BSY line by the external device.

Equal-priority polled eight-bit I/O

Let's consider, finally, the equal priority case in Table 5-3. Frankly, this connection is of somewhat limited value; but it is an interesting exercise to see just what can be done.

There are eight input lines and eight output lines, so there is full-byte bidirectional data flow. Both are basically blind links. In some applications this might be unimportant. If the output lines from the PC are being used for static control and the input lines to the PC represent ongoing status information that is being monitored, the data is essentially always valid—at any moment it reflects the current status of the whole setup.

There is one line left over, which could perhaps be used for signaling. Suppose you choose the *STB line as the lone signal line. The *STB line can be used as either an input or an output, which suggests sharing of the line by both devices if some form of take-turns (half-duplex) data exchange is acceptable. Such a connection is depicted in Fig. 5-16. The pin assignments have been selected so as to require the least amount of bit shuffling by software.

In order for this scheme to work, both the PC and the external device must be able to release the *STB line and the *STB line must be pulled up so that it goes high when it is released. The open-collector drivers used for the *STB line in the PC parallel port allow this. (If the other device isn't another PC parallel port it must also use an open-collector or three-state driver for the *STB line.) The *STB line will be high if neither side is set to low. If either goes low, or both go low, *STB will be low. But if each side keeps a flag variable reflecting how it has set its own driver it will know whether it, or the other side, has brought *STB low. The pseudocode in Fig. 5-17 shows how each side can distinguish among the possible line states.

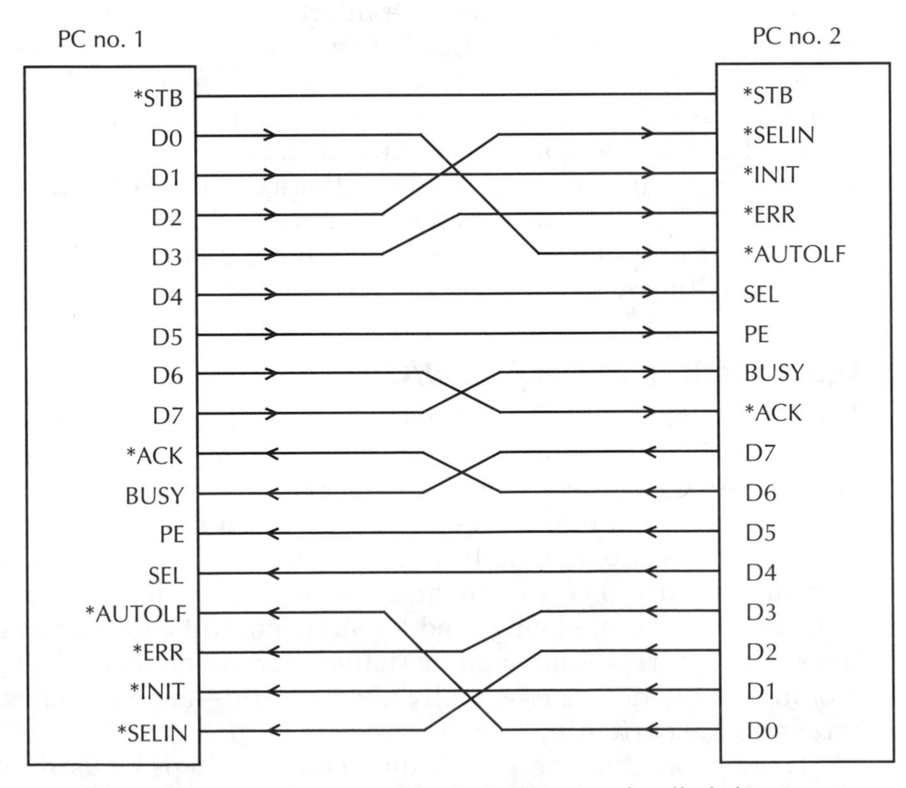

Fig. 5-16 *Connections for bidirectional polled I/O.*

Fig. 5-17 *Pseudocode for routines for sharing the *STB line.*

```
procedure AssertStb
begin
   MyStbOn := TRUE
   write 5 to port PCR to make driver output low
end

procedure ReleaseStb
begin
   MyStbOn := FALSE
   write 4 to port PCR to make driver output high
end

function StbIsLow
begin
   read port PCR
   if (value and 1 <> 0) return TRUE   { *STB is low }
      else return FALSE
end

Line States:
   Released if (MyStbOn = false) and (StbIsLow = false)
   Seized if (MyStbOn = false) and (StbIsLow = true)
   InUseByMe if (MyStbOn = true)
```

One possible way to implement half-duplex transmission is outlined in Fig. 5-18. The state of the *STB line is tested by reading the control port. If bit 0 is 1 then *STB is low and vice versa. The receiver detects valid data by temporarily releasing *STB and testing if it goes high. If so, it reads the data and then seizes *STB once again. An interesting possibility is to combine the *STB state test and the reading of PCR data in one operation. The bit shuffling required for reading data shifts the PCR byte one place to the right. Such a shift very conveniently puts bit 0 into the CPU carry flag. The function **PollSTB** in Fig. 5-19 uses such a ploy. If *STB is high, indicating valid data, the routine also reads the PSR and adds it in to give a total of eight data bits. (Although written in assembly language it is a Pascal subroutine. It must be linked as a far call.) The function returns a word with the upper byte zero if *STB is low, and with the FF hex in the upper byte and data in the lower byte if *STB is high. This allows a simple test: if the value returned by **PollSTB** is greater than 255 then the low byte has data; if not, *STB was still low and there is no data ready. Note that **PollSTB** uses a variable *ParCR* in the static data segment to get the address of the PCR; *ParCR* should be set to (base + 2).

Fig. 5-18 Example of a half-duplex protocol using a shared *STB line.

```
When sender has byte to transmit:
   1. tests for *STB high = ok to send
   2. if ok, then
      2a. writes data to output
      2b. releases *STB
      2c. checks for read acknowledgment by
         3a. repeatedly tests *STB until it is high
         3b. repeatedly tests *STB until it is low
         3c. seizes *STB
   4. sender now ready for next byte

Receiver polls for input by:
   1. tests for data ready by
      1a. release *STB
      1b. if *STB stays low then seizes *STB again and loops to 1a
          else
          2a. reads input
          2b. seizes *STB
```

Polled input with handshake The bidirectional data transmission just considered is rather awkward and ungainly and is no doubt more of theoretical interest than of practical application. But the receive processing routines used in it can be adapted to set up a simple method of full-byte polled input through the parallel port. The PC receives eight-bit data from an external device through the (shuffled) parallel inputs, and the *STB line is used in only one direction as a means to signal the PC that data is ready. One of the PC's parallel output lines is used for acknowledgment, forming a handshaking link that provides flow control and ensures that no data is lost. The remaining output lines could be used for various control functions in the external device.

This method is attractive also when the external device is not another PC parallel port because of the simplicity of the interface. The basic elements are shown in Fig. 5-20. The eight-bit data is sent to the PC through a buffer such as an LS244 or HC244. A positive edge-triggered D flip-flop (LS74 or HC74) drives the *STB line. When the external device has data ready, it applies a short pulse (1 μs is plenty) to the set pin of the flip-flop, raising the *STB line connected to its Q output to high level. The D input of the LS or HC74 is strapped low so that resetting can be done by a low-to-high transition on the acknowledgment line which is connected to the CLK input of the LS or HC74. The –Q output of the LS or HC74 could be used as a "wait" signal by the other circuits in the external device; data transfer should pause until the –Q output goes high again, indicating that the PC has taken the current data.

Fig. 5-19 *Routine to test the status of the *STB line.*

```
;-------------------------------------------------------------------
;    Routine for bidirectional data I/O through parallel port
;-------------------------------------------------------------------
; POLLSTB.ASM 1.0 ©1992 JHJ

DATA    SEGMENT PUBLIC BYTE

        EXTRN ParCR : word              ; parallel PCR port, Base+2

DATA ENDS

CODE    SEGMENT PUBLIC BYTE

        assume cs:code,ds:data

;-------------------------------------------------------------------
; function PollSTB : word;  { far }
; { tests *STB: if line was HIGH then data not ready; returns with AH=0 }
; { if line was LOW, returns with AH=FF and AL=data
; ------------------------------------------------------------------

        FlipMask equ 10000101b          ; mask to flip bits

        PUBLIC PollSTB

PollSTB PROC FAR
        push    bp
        mov     bp,sp
        xor     ax,ax                   ; preset zero return
        mov     dx,[ParCR]              ; PCR port
        in      al,dx
        shr     al,1                    ; carry gets bit 0, -STB
        jc      Done                    ; 1 = -STB was low (data not valid)
        and     al,7                    ; mask unused bits
        mov     ah,al                   ; store
        dec     dx                      ; set for PSR port
        in      al,dx
        and     al,0F8h                 ; mask unused bits
        or      al,ah                   ; add in PCR bits
        xor     al,FlipMask             ; flip so bit vals same as input lvls
        mov     ah,0FFh                 ; make AH nonzero
Done:
        mov     sp,bp
        pop     bp
        ret
PollSTB ENDP

CODE ENDS
END
```

Data transfer proceeds as follows. A polling loop in the PC tests the *STB line, and when it detects a high level it reads the data, and then when the data has been processed it sets an "acknowledge" line—data line D0 is used in this example—to low and then back to high to reset the flip-flop. The function **PollSTB** in Fig. 5-19 lends itself nicely to this application.

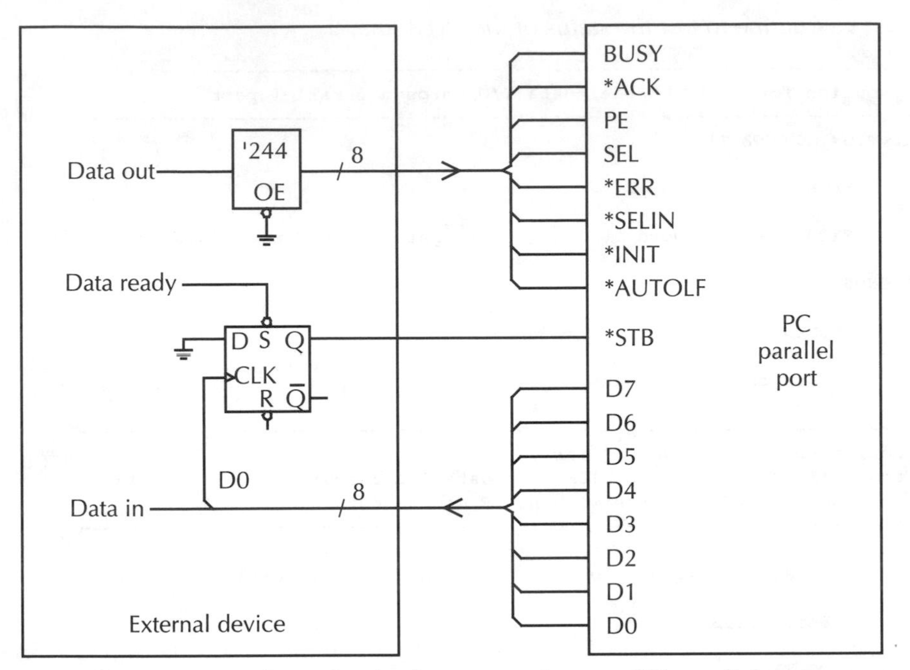

Fig. 5-20 *Interface circuits for connecting to a PC parallel port.*

Assuming that the external device has data ready as fast as the PC can take it, the maximum data transfer rate depends largely on how long the software in the PC takes to process incoming bytes. If they are simply stored in memory for later analysis, rather high rates are possible—10 to 15 kilobytes per second on an 8-MHz XT, and even faster on an AT or 386 machine. Because handshaking is used to control data flow, the actual rate might vary during operation. It will never be faster than whichever is the slower, the external device or the PC.

Interrupt-driven parallel input

In most applications interrupt-driven input operation provides superior performance. It is especially attractive when the parallel port is to be used primarily for receiving input, but it is also valuable even when the primary data flow is output from the port. With a receive buffer of adequate size, the occasional time-consuming task, such as writing a block of data to disk, need not slow down the rate of data flow. Because the *ACK line is used to trigger the interrupt, only eight lines are left over for output from the port. As suggested by Table 5-3, they can be used either as eight unstrobed general-purpose data lines, or one line can be used as a strobe line leaving seven data lines.

The principles underlying hardware ISRs and their implementation were discussed in chapter 4 and won't be repeated here.

As mentioned in the description of the standard parallel port adapter, the implementation of the interrupt request (IRQ) is flawed in the design of the original adapter. Some later designs have corrected this problem. Unfortunately, determining whether a port uses the old or a newer design is usually a matter of trial and error unless the documentation for the port says whether the *ACK line is latched. (If there is an option for leading-edge or trailing-edge operation it is a safe bet that *ACK is latched.) It is likely, though not guaranteed, that a built-in (motherboard) parallel port on an AT or 386 machine is of the newer sort. You cannot safely assume that all of the parallel ports in a newer machine are the improved version because many plug-in parallel port I/O cards still follow the original IBM design.

If the parallel port does latch the *ACK line, the external device connected to the computer can emit a conventional strobe pulse of 5- to 10-μs width when it has data ready for the PC. This will not work if the *ACK line is not latched.

If the port does not latch *ACK, then there are two methods that can be used for interrupt-driven input. The first method is very simple. A stretched or extended *ACK strobe pulse is used that is long enough to ensure that the interrupt has been recognized. On a 4.77-MHz PC such a stretched strobe would be on the order of 100 μs wide; 50 μs will usually suffice on an 8- or 10-MHz PC or XT. The second method uses a form of handshaking, rather than a strobe pulse. The external device is designed to pull the *ACK line low when data is ready and then wait. Because the IRQ line in the PC will then remain high, the interrupt will be triggered and control will pass to the ISR. At this point it is safe to release the IRQ line. The ISR code begins by activating a response line (one of the data lines) that the external device monitors and uses to reset *ACK back to high.

If operation with either a latched or an unlatched port is desired, a compromise can be used. An extended strobe can be used to accommodate unlatched ports. On a latched port this will just waste time until the *ACK strobe returns to high level, activating the interrupt. The compromise consists of providing the external device with a selectable strobe width, say 100 μs for use with unlatched ports and 10 μs for latched ports. The hardware and interface examples in chapter 6 take this compromise approach and they have worked well with a considerable range of computers, new and old.

Let's turn now to the ISR itself. An ISR for parallel port input is shown in Fig. 5-21. It gets the input and puts it into a ring buffer, adjusts the buffer write pointer, and increments a count variable. It is a lot like the serial ISR in chapter 4, except that getting the input is more complicated inasmuch as two ports must be read and some bit manipulations done. Because there is a degree of ambiguity in the I/O addresses of the parallel ports, the port address is fetched from memory rather than being hard-coded as an immediate operand. Even so, its performance isn't too bad. It takes about 27 percent longer to execute than the serial ISR (in chapter 4) does. Like the serial port ISR, it is written in assembly language for compactness and speed. The medium memory model is assumed, with multiple code segments and a single static data segment.

An alternate version of the ISR written in Pascal is shown in Fig. 5-22. You can use it rather than the assembler version with some sacrifice of speed. Whichever version you use, don't forget that these are far (segment:offset) routines and require the {$F+} compiler directive.

The ISR follows the principles for writing interrupt routines discussed in chapter 4. Its work is done in four steps. First, all CPU registers that will be modified are saved on the stack. The address of the data segment is obtained (it is plugged in when DOS loads the program) so that the data segment variables *ParSR* (the I/O address of the parallel port status register), *ParBuffer, ParBufWrPtr,* and *ParBufCount* can be accessed. Second, the incoming data byte is obtained by reading the status register and control register and manipulating bits using the same techniques as in the **ParPortIn** function in Fig. 5-12. Third, after the incoming data byte has been built it is put in the buffer. The write pointer *ParBufWrPtr* is incremented and ANDed with *ParBufEnd,* which wraps the pointer if necessary. The variable *ParBufCount* is incremented. Fourth, an end-of-interrupt command is sent to the interrupt controller, the saved CPU registers are restored, and an interrupt return instruction returns control to the code that was executing when the interrupt occurred.

The ISR is used only for input. Output would be handled by writing a byte to the PDR at the base address. The bits in the byte correspond to data lines D0–D7. If any external device strobing is needed, one of the data lines must be given over for the purpose and the strobe created by a series of writes to the PDR with the

Fig. 5-21 The parallel toolkit ISR.

```
;---------------------------------------------------------------
;
;       Interrupt Service Routines for parallel port
;
;---------------------------------------------------------------
; PARTKISR.ASM 2.0 ©1992 J H Johnson

DATA    SEGMENT BYTE PUBLIC

        EXTRN ParBufWrPtr : word
        EXTRN ParBuffer : byte
        EXTRN ParSR : word
        EXTRN ParBufCount : word

DATA ENDS

CODE    SEGMENT BYTE PUBLIC

        assume cs:code,ds:data

        PUBLIC ParPortISR

        ;---------------------------------------------------------------
        ; procedure ParPortISR; { far }
        ;---------------------------------------------------------------

; equates
        ParBufEnd       equ 4095        ; must agree with constant in unit!
        IntEOIport      equ 20h
        EOI             equ 20h
        PolarityMask    equ 10001011b   ; mask to flip bits
        ; pins flipped thus: | -11 | +12 | +13 | +15 | -17 | +16 | -14 | -1 |

ParPortISR      PROC FAR

        push    ax                              ; save CPU registers
        push    bx
        push    dx
        push    ds

        mov     ax,SEG DATA                     ; set up data seg addr
        mov     ds,ax

; get input and shuffle bits
        mov     dx,[ParSR]              ; Base+1 address - first read
        in      al,dx                   ; get data in AL

; pins as bits in AL now :      | 11 | 10(ACK) | 12 | 13 | 15 | - | - | - |


        ; "copy" bit 7 (pin 11) to bit 6 (pin 10) and shift left
        and     al,0B8h                 ; force bit 6 to 0, mask bits 0..2
        shl     al,1                    ; shift, carry flag gets old bit 7
        jnc     P1                      ; old bit 7 was 0, ok to continue
        or      al,80h                  ; old bit 7 was 1, so set bit 7

; pins as bits in AL now :      | old 11 | 12 | 13 | 15 | - | - | - | 0 |

P1:     mov     ah,al                   ; ah will be upper nibble of result
```

Fig. 5-21 Continued.

```
        inc     dx                      ; next port addr
        in      al,dx

; pins as bits in AL now :      | - | - | - | IRQ en | 17 | 16 | 14 | 1

        and     al,0Fh                  ; keep lower 4 bits only
        or      al,ah                   ; add in upper nibble

; pins as bits in AL now :      | 11 | 12 | 13 | 15 | 17 | 16 | 14 | 1 |

        xor     al,PolarityMask         ; flip bits for correct sense

; save result which is now in AL
        mov     bx,[ParBufWrPtr]        ; get buffer write pointer
        mov     [ParBuffer+bx],al       ; store input byte
        inc     bx
        and     bx,ParBufEnd            ; wrap if need be
        mov     [ParBufWrPtr],bx        ; new pointer value
        inc     WORD PTR [ParBufCount]

        mov     al,EOI                  ; reset PIC
        out     IntEOIport,al

        pop     ds                      ; restore registers
        pop     dx
        pop     bx
        pop     ax
        iret

ParPortISR      ENDP

CODE    ENDS

END
```

Fig. 5-22 *Pascal version of the parallel toolkit ISR.*

```
{ Pascal version of ISR in the Parallel Toolkit }

{ This routine must be declared as FAR }

{$F+ }

procedure ParPortISR;
interrupt;
var
   B : byte;
begin
   B := (Port[ParSR]);
   if B and $80 = 0 then B := (B and $38) shl 1
      else B := ((B or $40) shl 1) and $F0;
   ParBuffer[ParBufWrPtr] := (B or (Port[ParRegC] and $0F)) xor PolarityMask;
   ParBufWrPtr := (ParBufWrPtr+1) and ParBufEnd;
   inc( ParBufCount );
   Port[IntEOIport] := EOI;
end;

{$F- }
```

appropriate bit set or cleared. The procedure Strobe7 in Fig. 5-6 illustrates the process, using the D7 line as the strobe line.

As with all modifications of the interrupt vector table, a program that installs the parallel ISR must take care to restore the original interrupt vector before it terminates and exits back to DOS.

The parallel port toolkit

The Parallel Port Toolkit, shown in Fig. 5-23, provides the core routines for high-performance parallel data input. It provides interrupt-driven eight-bit input and is intended for applications in which the primary data path is input to the computer. Eight output lines remain available, which can be used for eight-bit blind or static output or configured as seven-bit data output with a data signal or strobe.

The toolkit illustrates the practical application of the principles already discussed in this chapter. It combines interrupt-driven receiving with polled transmission and status checking. As with the CRC Toolkit presented in chapter 3 and the Serial Toolkit in chapter 4, it is designed as a library of routines packaged in a module. The toolkit can be used as is in programs for parallel-port I/O or modified as needed to suit your particular needs. It is also offered for your study to see how the things discussed here are actually done in working, tested code.

Receive buffer and the ISR

Receive (input) operations are interrupt-driven. The ISR discussed in the previous section fetches each incoming byte from the parallel port, manipulates it to proper form, and puts it in a buffer. The main program gets parallel-port input by retrieving bytes from the receive buffer.

The receive buffer, *ParBuffer*, is declared explicitly and its last element (its size minus one) is declared as a constant, *ParBufEnd*. This has the (possible) disadvantage that its size cannot be changed during execution and that setting a different size entails recompiling the unit. On the other hand, it has the advantage that it allows the buffer pointer manipulations in the ISR to be written using immediate operands which produce smaller and faster code. The size shown in the listing is 4096 bytes, a good size for general-purpose use. If the incoming data rate is relatively low a smaller buffer would suffice. A larger buffer might

Fig. 5-23 The parallel toolkit unit.

```
{ ---------------------------------------------------------------------------- }
{                              PARALLEL TOOLKIT                                  }
{          A Library of Bidirectional Parallel Port I/O Routines                 }
{ ---------------------------------------------------------------------------- }
(* PARTK.PAS 1.1 ©1992 J H Johnson *)

{ ---------------------------------------------------------------------------- }
{                    Parallel Port Connector Pin Mapping                         }
{       input from external                         output from par port         }
{    Di7 pin 11        Di3 pin 17                  Do7 pin 9    Do3 pin 5        }
{    Di6 pin 12        Di2 pin 16                  Do6 pin 8    Do2 pin 4        }
{    Di5 pin 13        Di1 pin 14                  Do5 pin 7    Do1 pin 3        }
{    Di4 pin 15        Di0 pin  1                  Do4 pin 6    Do0 pin 2        }
{    -strobe pin 10                                                              }
{ ---------------------------------------------------------------------------- }

UNIT
   ParTk;

INTERFACE

USES
   Dos;

CONST
   { *** SET BUFFER SIZE HERE *** }
     { parallel buffer size is always 2^N }
     { last element is (2^N)-1, e.g. 255,511,1023,2047,4095,8191,16383,32767 }
   ParBufEnd = 4095;           { must be same as the equate in PARTKINT.ASM ! }

VAR
   ParBufCount : word;                  { overflow = (ParBufCount > ParBufEnd) }


function GetParBase( N : byte ) : word;
     { 1 = LPT1, 2=LPT2; returns port base address }

procedure OpenParPort( P : byte );
     { call with 1=LPT1 (uses IRQ7), 2=LPT2 (uses IRQ5) }
     { if using LPT2, be CERTAIN that machine uses IRQ5 for printer port #2 }

procedure CloseParPort;
     { close and restore original interrupt vector }

function ParOk : boolean;
     { true if port open and routines were installed successfully }

function ParDataWaiting : boolean;
     { true if data in buffer }

procedure FlushParBuffer;
     { empty the buffer }

function ReadPar : byte;
     { read input from external device }
     { bit mapping, msb..lsb: BSY,PE,SEL,ERR,SELIN,INIT,AUTOFD,STB }

function PeekPar: byte;
```

Fig. 5-23 Continued.

```
     { next byte in Rx buffer, 0 if buffer empty; byte remains in buffer }

function ReadPar7 : byte;
     { read 7-bit input from external device }
     { bit mapping, msb..lsb: (0),BSY,PE,SEL,ERR,SELIN,INIT,AUTOFD }

procedure WritePar( B : byte );
     { outputs D0..D7 (pins 2..9) assume value B (no strobe) }

IMPLEMENTATION
{ ================================================================================ }

{ ----- Cross-reference to Assembler declarations ----- }
{ These static variables are declared in PARTKINT.ASM:
  DATA SEGMENT BYTE PUBLIC
       EXTRN ParBufWrPtr : word
       EXTRN ParBuffer : byte
       EXTRN ParSR : word
       EXTRN ParBufCount : word
DATA ENDS
This equate MUST agree with the value in the unit: ParBufEnd equ 4095        }
{ ---------------------------------------------------------- }

CONST
   { for interrupt handler }
   IntMaskReg = $21;                                                    { PIC }
   Int5mask = $20;                                                { IRQ5 mask }
   Int7mask = $80;                                                { IRQ7 mask }
   Lpt2Int = $0D;                                               { IRQ5 intr nr }
   Lpt1Int = $0F;                                               { IRQ7 intr nr }

VAR
   ParBase,                               { PDR (data) at port base address }
   ParSR,                                         { PSR (status), base + 1 }
   ParCR : word;                                 { PCR (control), base + 2 }
   ParBuffer : array [0..ParBufEnd] of byte;
   ParBufWrPtr,ParBufRdPtr : word;              { input ring buffer pointers }
   OldParIntVec : pointer;
   IntMask : byte;
   ParInt : byte;
   Device : byte;              { 1=LPT1, 2=LPT2; any other = port not opened }
   ExitSave : pointer;

{$F+ }

{$L PARTKINT.OBJ }
procedure ParPortISR; external;

procedure ParExit;
begin
   CloseParPort;
   ExitProc := ExitSave;
end;

{$F- }

function GetParBase( N : byte ) : word;                       { 1 = LPT1, etc. }
```

Fig. 5-23 Continued.

```
var
   Table : array [0..3] of word absolute $40:0008;     { BIOS port addr table }
begin
   N := (N-1) and 3;                                      { adjust N and clamp }
   GetParBase := Table[N];                                         { get value }
end;

procedure OpenParPort( P : byte );
var
   Mask : byte;
   A : integer;
begin
   if (Device <> 0) or (P<1) or (P>2) then Exit;
   A := GetParBase( P );
   if A = 0 then Exit;
   { set up addresses }
   ParBase := A;
   ParSR := A+1;
   ParCR := A+2;
   { set up for LPT1 or LPT2 }
   if P = 2 then
      begin
         IntMask := Int5mask;                                { to enable IRQ5 }
         ParInt := Lpt2int;
      end else
      begin
         IntMask := Int7mask;                                { to enable IRQ7 }
         ParInt := Lpt1int;
      end;
   { set up vectors }
   GetIntVec( ParInt,OldParIntVec );
   SetIntVec( ParInt,@ParPortISR );
   { set up PIC }
   Mask := Port[IntMaskReg];
   Port[IntMaskReg] := Mask and (not IntMask);

   { preset Base+2 bits and enable port IRQ }
   Port[ParCR] := $14;
   Device := P;
end;

procedure CloseParPort;
var
   Mask : word;
begin
   if (Device = 0) then Exit;
   Port[ParCR] := $04;                { reset Base+2 bits and disable IRQ }
   Mask := Port[IntMaskReg];                                  { restore PIC }
   Port[IntMaskReg] := Mask or IntMask;
   SetIntVec( ParInt,OldParIntVec );                       { restore vector }
   Device := 0;
end;

function ParOk : boolean;
begin
   ParOk := (Device = 1) or (Device = 2);
```

Fig. 5-23 Continued.

```
end;

function ParDataWaiting : boolean;
begin
   ParDataWaiting := (ParBufRdPtr <> ParBufWrPtr);
end;

procedure FlushParBuffer;
begin
   ParBufRdPtr := 0;
   ParBufWrPtr := 0;
   ParBufCount := 0;
end;

function ReadPar : byte;
begin
   if ParBufRdPtr <> ParBufWrPtr then
      begin
         Inline($FA);                          { CLI to disable interrupts }
         ReadPar := ParBuffer[ParBufRdPtr];
         ParBufRdPtr := succ(ParBufRdPtr) and ParBufEnd;
         Inline($FB);                                { STI to restore ints }
         dec( ParBufCount );
      end else ReadPar := 0;
end;

function ReadPar7 : byte;
begin
   if ParBufRdPtr <> ParBufWrPtr then
      begin
         Inline($FA);                          { CLI to disable interrupts }
         ReadPar7 := ParBuffer[ParBufRdPtr] shr 1;     { shift right 1 digit }
         ParBufRdPtr := succ(ParBufRdPtr) and ParBufEnd;
         Inline($FB);                                { STI to restore ints }
         dec( ParBufCount );
      end else ReadPar7 := 0;
end;

function PeekPar : byte;
begin
   if ParBufRdPtr <> ParBufWrPtr then
      PeekPar := ParBuffer[ParBufRdPtr]
         else PeekPar := 0;
end;

procedure WritePar( B : byte );
begin
   Port[ParBase] := B;
end;


BEGIN
   ExitSave := ExitProc;
   ExitProc := @ParExit;
   Device := 0;
   FlushParBuffer;
END.
```

be necessary if the main program contains time-consuming routines such as real-time graphics and fairly high data rates are being used.

ParBuffer is arranged as a ring buffer using read and write pointers *ParBufRdPtr* and *ParBufWrPtr*. The same technique is used for wrapping around the end of the buffer as was used in the Serial Toolkit in chapter 4. The buffer size is constrained to be a power of 2 so that it consists of 2^N elements, 0 to $2^N - 1$. Because $2^N - 1$ will always be all 1s, the pointer can be wrapped simply by ANDing it with $2^N - 1$. This is much faster than comparing the pointer to a maximum value and setting it to zero if greater.

If you want to use a different receive buffer size, remember that the fast wrap ploy permits only certain sizes. The value specified for *ParBufEnd* must be some $2^N - 1$; for instance, 63, 127, 255, 511, 1023, 2047, 4095, 8191, 16,383, or 32,767. (The corresponding hexadecimal values are 3F, 7F, FF, 1FF, 3FF, 7FF, FFF, 1FFF, 3FFF, 7FFF.) **Important:** If you change buffer size and you are using the assembly language ISR, you must change the *ParBufEnd* equate in the .ASM file also!

The *ParBufCount* variable is incremented each time the ISR executes, and it is decremented each time a byte is retrieved from the buffer. *ParBufCount* is included for two reasons. One is to allow an overflow of *ParBuffer* to be detected. Ordinarily, the long-term average value of *ParBufCount* will be zero because each increment of its value by the ISR is offset by a read from the buffer. If the buffer overflows, the reads start over again at zero so *ParBufCount* will not be fully decremented. Overflow is indicated if *ParBufCount* is greater than *ParBufEnd*. The other purpose of *ParBufCount* is to allow buffer usage to be monitored. If its value is read immediately after a time-consuming routine, such as writing to a floppy disk, you can get an idea of how many bytes have accumulated in the meantime. This can be helpful in selecting a reasonable buffer size.

Control functions

OpenParPort() and **CloseParPort** open and close the parallel I/O channel, much as you would open or close a disk file. They encapsulate the installation and removal of the ISR and IRQ. **OpenParPort** is called with the parameter 1 or 2 to select LPT1 or LPT2, respectively, to be the active port. **CloseParPort** restores the original interrupt and must be called at some point before the main program terminates and exits to DOS. (As a precaution, the exit routine of this Unit calls it.)

OpenParPort opens LPT1 using interrupt request line IRQ7 which is standard in all PCs. LPT2 is opened using IRQ5. This is usually valid in AT and 386 machines, although it is wise to check the computer's documentation to be sure. Do not try to open LPT2 on an XT—IRQ5 is used for the hard disk in these computers! When an LPT2 port is installed in an XT it is often assigned to IRQ2. The toolkit can easily be modified to use IRQ2 for LPT2 by doing the following:

- Declare an additional constant Int2mask = $04.
- Change the Lpt2int constant to Lpt2int = $0A.
- Change all Int5mask references in the procedure **OpenParPort** to Int2mask.

GetParBase is included because of the uncertainty as to which parallel port is at which I/O address. It is called with 1, 2, or 3 for LPT1, LPT2, or LPT3, and it returns the base address for that device from the BIOS table. A return of 0 indicates that there is no hardware indicated for that LPT device.

FlushParBuffer clears the receive buffer. It is a convenient way to discard unwanted information or garbage. It is a good idea to call it after opening the parallel channel to make sure that the buffer is empty.

Status functions

ParDataWaiting returns true if the receive buffer is not empty; that is, there are one or more characters in the receive buffer waiting to be read. **ParOk** returns TRUE if the parallel channel is open, or FALSE if it is not.

I/O functions

ReadPar reads and removes a byte from the receive buffer. If the buffer is empty it returns 0. It is the normal read function used for eight-bit input. The mapping of the bits of the byte to the parallel port lines is the same as used earlier in this chapter for eight-bit input routines and is also shown in the interface section of the toolkit listing for reference.

PeekPar reads a byte from the buffer but does not remove it. If the buffer is empty it returns 0. It provides a one-byte look ahead that is sometimes useful. The bit assignments are the same as in the **ReadPar** function.

ReadPar7 reads and removes a byte from the receive buffer and returns it as a seven-bit value. If the buffer is empty it returns

0. It is a special read function for use when the *STB line is being used as an output strobe. The mapping of input lines to bits is similar to that for eight-bit input with the following differences: (1) the *STB line is not used; and (2) all of the bits in the byte are shifted one place to the right.

WritePar writes a byte to the parallel port data register, setting output lines D0 through D7.

In some applications it might be desirable to combine the interrupt-driven input provided by the parallel toolkit module with strobed output. Because only eight output lines are available, only seven-bit data can be transmitted. The procedure **WriteStrobe()** in Fig. 5-24 illustrates a simple way to do it. Output line D7 is used for a negative-going strobe pulse and output lines D0–D6 contain the seven-bit data.

Fig. 5-24 *Routine for strobed output.*

```
{ Routine to strobe the parallel port output line D7. }

procedure WriteStrobe( D : byte );
begin
   WritePar( D or $80 );                     { write data with D7 high }
   WritePar( D and $7F );                      { set D7 low to strobe }
   WritePar( D or $80 );                            { restore D7 high }
end;
```

Using the parallel port

The Parallel Port Toolkit was purposely written to be similar in structure and operation to the Serial Toolkit presented in chapter 4 in order to make it as easy as possible to use either serial or parallel I/O in application programs. As a matter of fact, the concept of a "communication engine" and the processes of flow control discussed in chapter 4 for serial data transmission apply equally to the parallel port, although most people don't ordinarily think of these topics in connection with parallel data interchange. If a parallel interface is used with the digital voltmeter presented in chapter 6, the only changes needed in the sample data acquisition software in chapters 6 and 7 is the substitution of the parallel toolkit.

The chances are that if you use the parallel port it will be mainly for data I/O to external devices such as sensors and other instruments. Chapter 6 presents a number of I/O interfaces for use with the parallel port. One major advantage of using parallel transmission is that the interface hardware in the external device is simpler and less expensive. A couple of bus buffers, a flip-flop, and a few NAND gates suffice. A serial interface would need a UART, a baud rate generator, and RS-232 drivers.

Of course, the parallel port doesn't have to be used for conventional data at all. Another way to look at it is as a group of seventeen TTL circuits, some of which are for output, some for input, and some of which can be used either way. These lines can be used individually or in various groups to control external devices, such as relays or solenoids or LEDs, or to receive various kinds of on-off or status inputs from external devices. The possibilities are nearly endless.

The previous paragraph makes a point, implicitly, that is well worth making and which is one of the themes of this book. The point is this: that it is good to have a certain looseness or fluidity in how you think of a computer and its uses. There is nothing wrong with thinking of computers and other things in conventional ways. On the contrary, it is necessary if you are not to waste a lot of time constantly reinventing the wheel. But there are also times when it is better to let the mind drift a little and to consider unconventional uses. The key to success in these forays into the unconventional is a thorough understanding of the basic principles. That is why they have been stressed in this book. Thinking of the printer data output lines as eight wires connected to a TTL buffer might suggest all sorts of useful applications. The serial port modem control lines might likewise be used for all sorts of things. And if an idea doesn't work, don't throw it away until you are sure that you understand why it doesn't and where you went wrong. This is not merely for learning how not to make the same mistake in the future. It is also because today's flop, properly understood, might suggest tomorrow's really good bright idea.

Summary

This chapter has investigated ways to use the standard PC printer ports for general parallel input and output. The task is complicated by various quirks in the design of the parallel port hardware, and ways were considered to work around them.

The techniques developed for using the parallel port for I/O were brought together in a library module, the Parallel Port Toolkit. Like its counterpart in chapter 4, the Serial Toolkit, its purpose is to handle the low-level work of input and output and control of the port. The design of the two toolkits was purposely similar to facilitate using either one in an application program.

At this point the operation and programming of the serial and parallel ports, the I/O operations in the PC itself, and the basic principles involved in incorporating I/O functions in programs have been discussed. The chapter discusses a variety of external devices that can be used with any PC to construct a wide range of data acquisition and control systems.

❖6 Hardware examples

The preceding chapters have been concerned mainly with getting data into and out of a PC. But the data have to come from somewhere! So now let's look at some hardware for interfacing the computer with the "real" analog world. This chapter presents a collection of hardware modules—circuit building blocks that can be combined to create many different and versatile systems. One of them is more accurately termed a subsystem: a versatile 4½-digit dc voltmeter module with performance equal to that of commercial instruments.

The circuits described in this chapter use standard analog and digital ICs, with some readily available special-function chips where appropriate. The digital parts are complementary metal-oxide semiconductor (CMOS) devices that provide higher noise immunity than bipolar (LS-TTL) circuits and very low power consumption. (The latter makes battery operation feasible which can be very useful in certain situations.) If necessary, bipolar TTL devices can be substituted for their CMOS counterparts except in a few places where the very high input impedance of CMOS is used to advantage. Chips that can be substituted across families are indicated by an apostrophe in the type number: for example, a '393 could be either a 74HC393 CMOS part (recommended) or a bipolar 74LS393. There is a minor caveat: the logic levels of CMOS and TTL devices are not the same. The output of lightly loaded CMOS devices is essentially either 0 V or +5 V, which is fine for driving TTL. On the other hand, the input switching threshold for TTL is nominally 1.2 V, whereas CMOS switches at about half the supply voltage; that is, 2.5 V. The high-level output voltage of most TTL circuits is poorly defined and is about 2.4 to 3 V—marginal for driving CMOS. This can be fixed by connecting a pull-up resistor (10 kΩ is fine in most cases) between the TTL output and +5 V. The compatibilities of combinations of 74LS TTL, 74HC CMOS, and the older and slower CD4000-series CMOS parts are summarized in Table 6-1.

Table 6-1 Logic compatibility.

Device families	Compatibility (+5-V supply voltage)
74LS and 74HCT	Full
CD4000, 74HC, and 74HCT	Full
74HC driving 74LS	Full
74LS driving 74HC	Use pull-up (e.g., 10 kΩ)
74LS driving CD4000	Use pull-up (e.g., 10 kΩ)
CD4000 driving 74LS	1 LS input load maximum

The 74HCT series is an exception. It is HC-CMOS with special input circuitry to make its input logic levels similar to LS-TTL. The 74HCT is available primarily in bus buffer and I/O types that might need to interface with other equipment using TTL devices.

Although good construction practices should be followed, none of the building blocks described in this chapter is unduly critical as to layout, nor are special construction techniques such as ground planes or multilayer circuit boards required. (In some cases, such as A/D and D/A circuits, special care must be taken with grounding. The proper connections are shown in the associated schematic diagrams.) Various circuit modules using these building blocks were built on standard single-sided 4½-by-6 inch plug-in perforated boards (Vero or Vector) using ordinary IC sockets and 28 AWG hookup wire for interconnections. Please note that wire-wrap techniques, while convenient, are unsatisfactory for analog circuitry. Remember that all unused inputs on logic devices should be connected to ground or +5 V. This is necessary with CMOS to avoid excessive device dissipation, and highly desirable with TTL to avoid noise pickup.

Parallel interface circuits

Because the hardware modules will usually be connected to a PC for data interchange and control it is reasonable to begin with I/O interface circuits. First, several circuits for parallel I/O using the parallel port as discussed in chapter 5 are considered.

Parallel data output to a computer

A circuit for parallel data output is shown in Fig. 6-1. A 74HC244 (or 74LS244) buffer drives the output lines. The 22-Ω resistors improve the impedance match between the buffer and the cable, a topic discussed in chapter 7. Many commercial products omit such resistors (whether from ignorance or to save a few pennies I do not know).

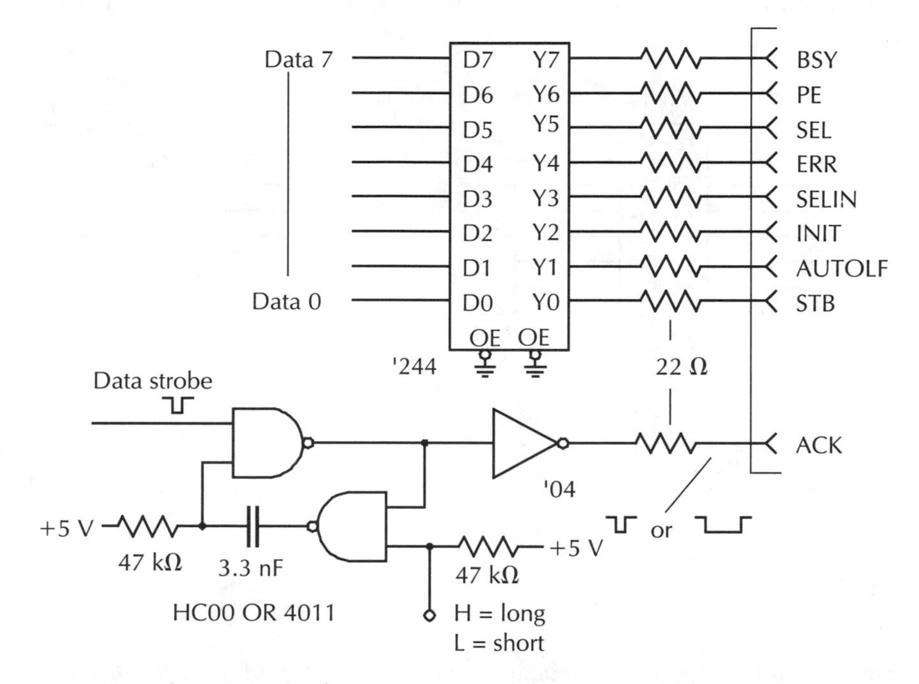

Fig. 6-1 *Interface to transmit data to a PC parallel port.*

The data strobe input should be a negative-going pulse of about 2- to 10-μs width. As discussed in chapter 5, the original PC parallel port does not latch the incoming interrupt strobe. To deal with this problem, the parallel block includes a programmable pulse stretcher. When Pulse Stretch is at high level, the strobe line to the PC is kept low for a long enough period to cover the interrupt latency in the PC, then goes high. With the component values shown in Fig. 6-1, the strobe to the PC is extended to about 110 μs. When used with the newer PC parallel ports that do handle the interrupt strobe correctly, Pulse Stretch is pulled to low level, and the data strobe is sent on to the PC as is. The NAND gate used for the pulse stretcher must be a CMOS type such as 74HC00 or CD4011. The width of the stretched output pulse is determined by the time constant of the RC network on one input to the first NAND gate, and is approximately 0.7 RC. The values shown, 47 kΩ and 3.3 nF, provide a stretched width of about 110 μs.

A companion input circuit block is shown in Fig. 6-2. Version A provides a strobe input, but as discussed in chapter 5 this leaves only seven lines for data. The data strobe latches the data in the '373 octal latch and also sets a '74 flip-flop to provide a "data ready" signal. Version B provides eight input lines without a data strobe.

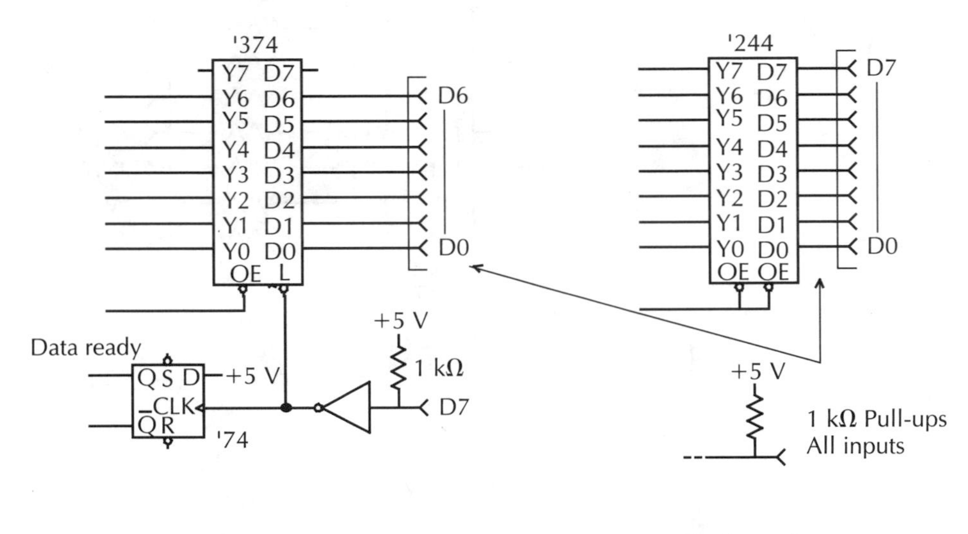

Fig. 6-2 *Interface to receive data from a PC parallel port, (a) with strobe and (b) without strobe.*

Parallel data input from a computer

Very much the same arrangement can be used when data flow is primarily from the PC to the external device (see Fig. 6-3). Data is received from the PC into a '373 latch. The device can send back to the PC in either of two formats: eight lines without strobe or seven lines with strobe. The parallel port *ACK line is used for data in the former case and as an interrupt strobe in the latter. The pulse stretcher from Fig. 6-1 can be used for strobed operation.

Printer emulation

Yet another variation on parallel I/O is to design the interface circuit to emulate a printer. Ordinary printer port routines cannot be

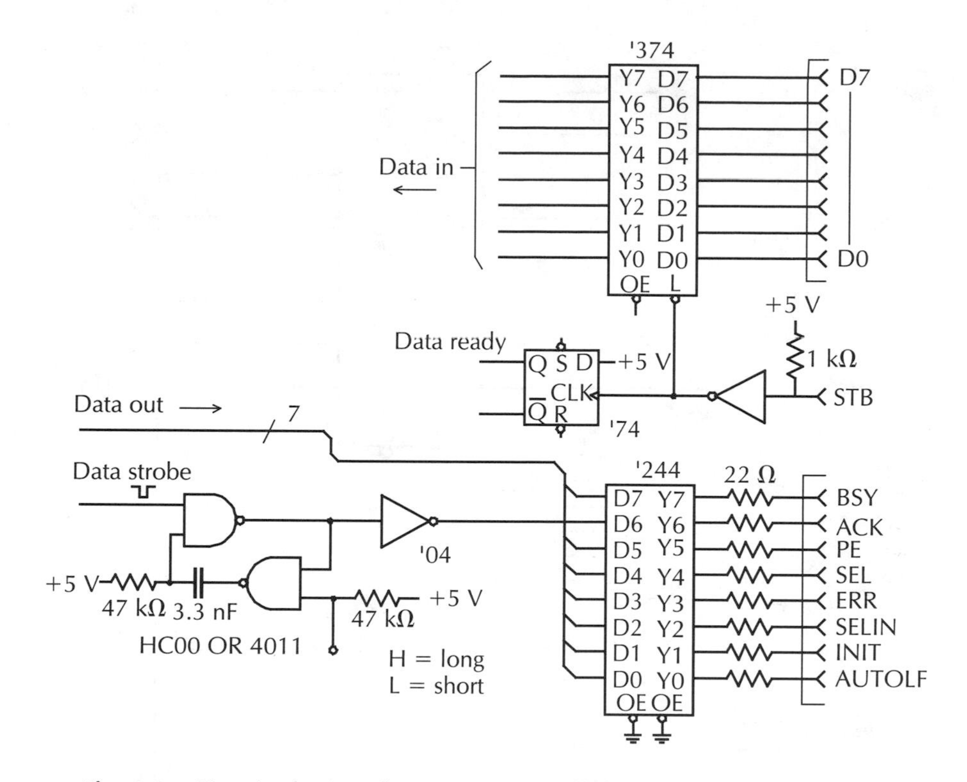

Fig. 6-3 *Circuits for interface to a PC parallel port with strobed receive.*

used to send out data from the PC to the parallel interfaces described previously because the BIOS checks the parallel port status register each time a character is sent and interprets whatever it finds as printer status information. One way to get around this is to have the external device act like a printer. This is done in the hardware in Fig. 6-4 which is otherwise similar to the previous examples.

For normal operation, the status lines must act like those of a printer: namely, PE = 0, *SEL = 1, *ERR = 1, and *ACK = 1 (or active-low acknowledge strobe). The BSY line handles flow control: 0 = go, 1 = wait. It is connected to the input latch flip-flop. The circuits in the external device reset the flip-flop when ready for the next byte.

The following usage of the status lines is suggested:

- SEL: 1 = okay, 0 = not ready or not in remote mode
- ERR: 1 = okay, 0 = serious or fatal error
- PE: 0 = okay, 1 = pause or trivial error

The circuit in Fig. 6-4 puts an additional inverter on the PE output so that a high level always means okay.

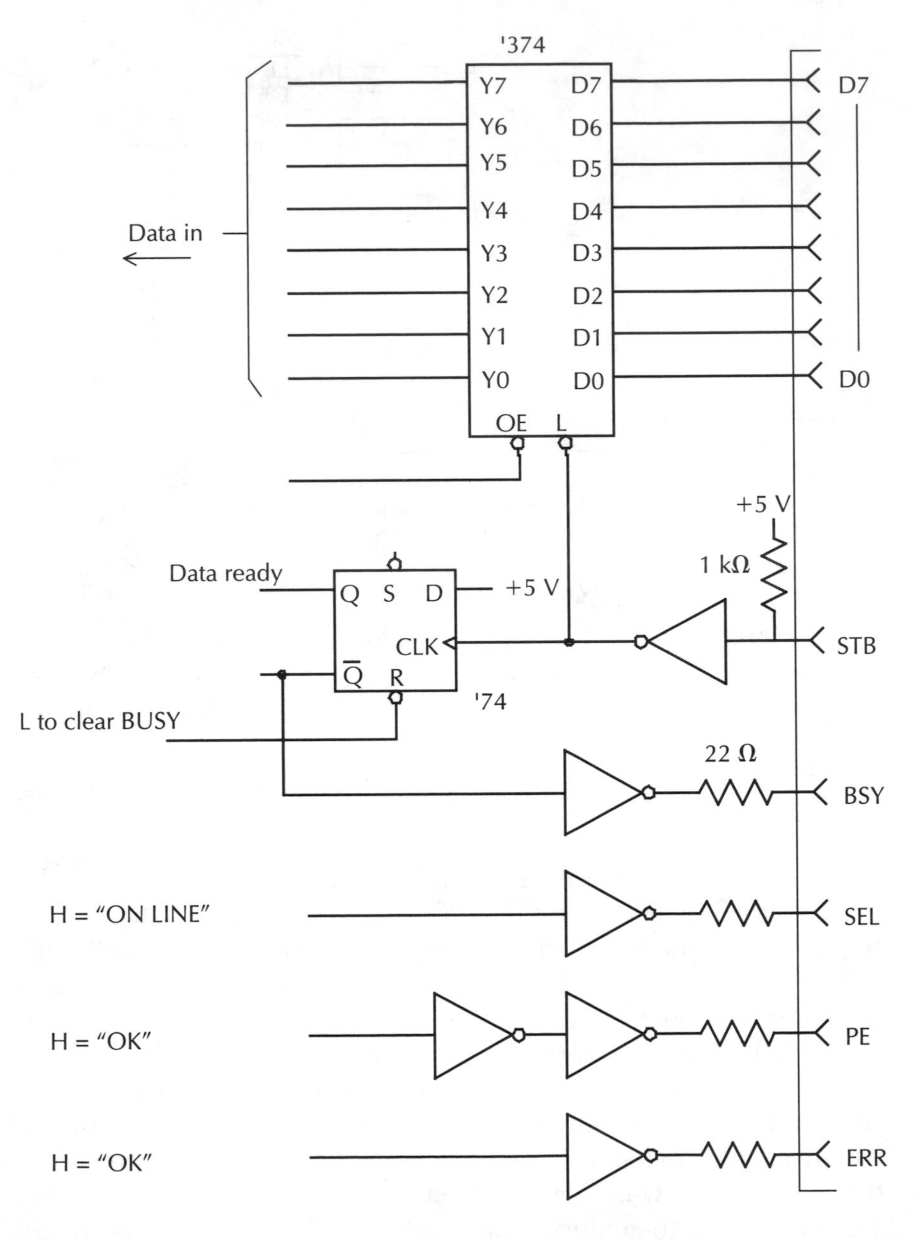

Fig. 6-4 *Circuits for emulating a printer.*

Serial interface circuits

A serial interface has the advantage of simpler cabling and also the ability to use such transmission paths as a modem and telephone line or a ham radio channel. If optical couplers are used for complete ground isolation, only one coupler for each path (transmit or receive) is required. Against these advantages must be weighed the somewhat more complicated interface with its UART, clock generator, and line drivers.

The serial I/O circuit block is built around the Harris-Intersil 6402, a CMOS implementation of an industry standard UART design. The pinout is shown in Fig. 6-5. All inputs are directly compatible with LS-TTL as well as CMOS. The functions of the inputs and outputs are as follows:

- TBR1–TBR8, RBR1–RBR8—the transmitter and receiver buffer input lines respectively. Five- to seven-bit operation uses only lines 1 through 5 to 7.
- RRC and TRC—the receiver and transmitter clocks respectively, which are sixteen times the baud rate. They are usually connected to the same clock source, but can be operated from different clocks if transmit and receive need to be at different baud rates. The 6402 does not contain a programmable rate divider.
- RRI and TRO are the serial input and output lines. Their polarity is high (5 V) for mark and low (0 V) for space.

The 6402 contains a control register that holds the transmission parameters selected by five configuration inputs. A high level on CRL loads the control inputs into the control register. When the transmission parameters are hard-wired or set by switches, CRL is simply tied to +5 V.

Control inputs

CLS1 and CLS2 set the word length, five to eight bits. SBS sets the number of stop bits: low for one and high for two bits (or 1½ bits if set for five-bit words). A high level on PI inhibits parity. A high level on EPE selects even parity and a low level selects odd. Table 6-2 shows the combinations for word length (number of data bits) and parity selection. An x indicates "don't care." When only seven- and eight-bit operation is required, as is usually the case, CLS2 can be strapped permanently high and CLS1 then selects between seven bits (low) or eight bits (high).

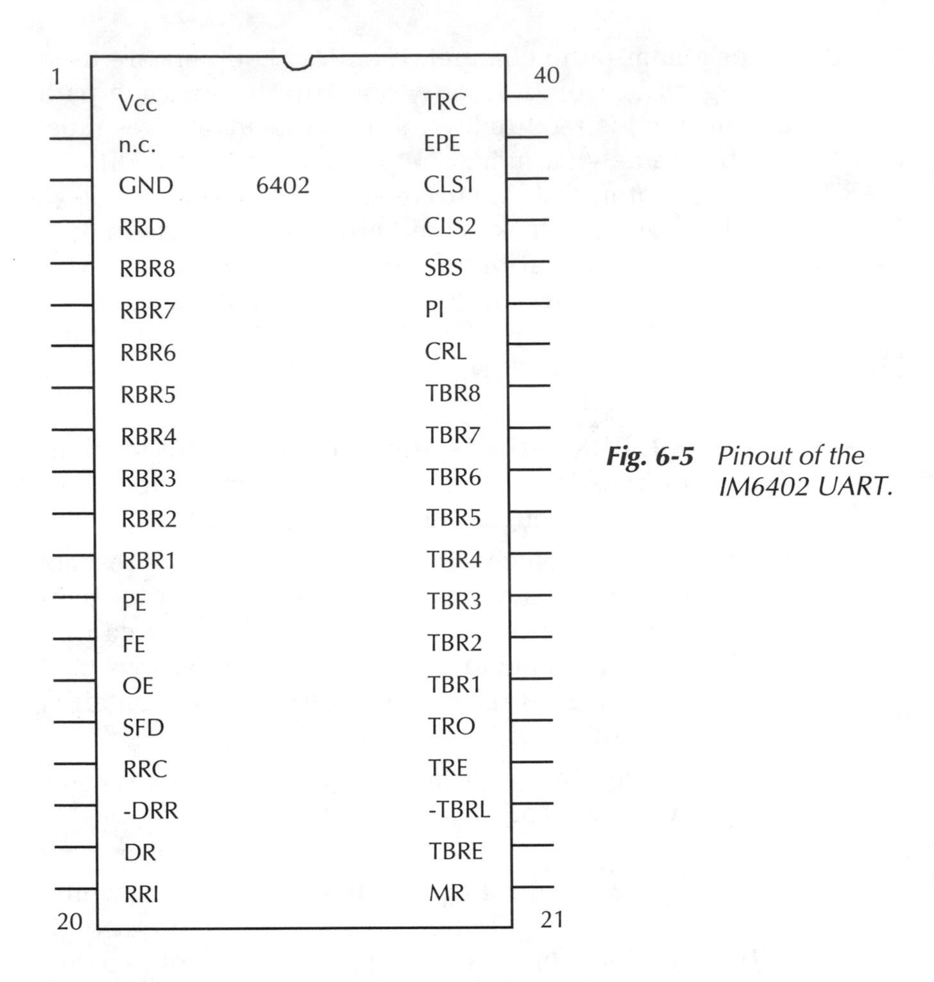

Fig. 6-5 *Pinout of the IM6402 UART.*

Table 6-2 Setting 6402 transmission parameters.

Word length, bits	**CLS1**	**CLS2**
5	L	L
6	H	L
7	L	H
8	H	H
Parity	**PI**	**EPE**
None	H	x
Even	L	H
Odd	L	L

The transmitter buffer input strobe line (–TBRL) is active low. The data on TBR1 to TBR8 are latched on the rising edge of the strobe.

TBRE is high when the transmitter buffer is empty and ready to accept the next character. TBRE goes low within one clock after data has been strobed into the transmitter buffer by the low-to-high transition of –TBRL.

TRE is high when the transmitter shift register is empty. As long as TRE is low, a character is still in the process of being transmitted.

DR is high when a character has been received and is ready to be read from the receiver buffer. DR remains high until –DRR (data received reset) is taken low. If DR is still high when the next character is received, an overflow will be flagged. In essence, –DRR is an acknowledge strobe that tells the 6402 that the current character has been read.

Receiver error flags

A high level on the OE pin indicates an overrun. Framing error (FE) is set high in the event of an error. If parity is enabled, PE goes high if there is a parity error. These outputs do not interrupt receiver operation and reset themselves when the error no longer exists.

The data and status outputs on the 6402 are three-state outputs which facilitates connecting it to a bus. A high level on RRD sets RBR1 through RBR8 to high impedance. A high level on SFD disconnects the error lines OE, FE, and PE and also DR and TBRE. When the 6402 is not sharing a bus with other devices, RRD and SFD are simply connected to ground so that the outputs remain enabled.

A high level on MR resets the UART, and is required at power-up. MR does not clear the receiver buffer.

UART clock circuits

The 6402 requires transmit and receive clocks at sixteen times the desired baud rate. The standard 6402 will operate with clocks up to 1 MHz (64 kbaud) and the 6402A will go up to 4 MHz (250 kbaud). The examples given in this chapter are for the standard 6402, which runs very nicely at 38,400 baud using a 614,400-Hz clock. The serial port routines in chapter 4 are designed expressly to handle high data rates.

The 6402, unlike the 8250 used in the PC serial port, does not contain a baud rate divider. Several baud rate generator ICs are available, but it is about as simple to use an ordinary binary counter chain as shown in Fig. 6-6. The input clock to the counters is 1.2288 MHz. A '93 four-bit binary counter provides UART

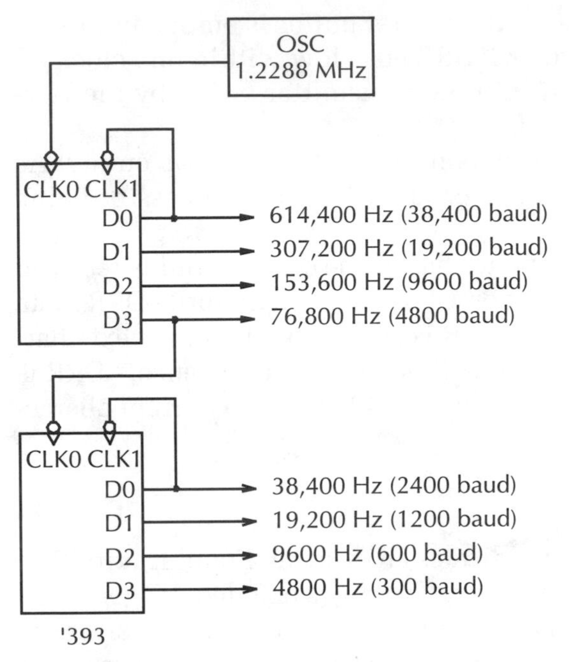

Fig. 6-6 *Baud clock divider circuit.*

clocks for baud rates of 38,400 down to 4800, which should be adequate for many applications. A second counter (or a '393 dual four-bit counter) extends the range to 300 baud. The chain could be extended even farther if very low baud rates are needed.

The divider chain can be driven by a packaged 1.2288-MHz oscillator, or a crystal oscillator circuit such as the one in Fig. 6-7a can be used. The inverters should be CMOS types such as the 74HC04 or the CD4049. The 74HC series NAND or NOR gates can also be used. The 33-pF pi-network capacitors together with typical stray capacitances are reasonable for most crystals. A divider chain can be used to accommodate other oscillator frequencies. Figure 6-7b shows a simple divide-by-four chain for use with a 4.9152-MHz crystal.

Line interface circuits

Four useful line interface circuits are shown in Fig. 6-8. Implementation is simplified by the fact that standard ready-to-use ICs are available for all circuits but one, the current loop. Recommended IC part numbers are indicated in the drawing.

Circuit A is the standard RS-232 interface. The modem control lines are omitted; as pointed out in chapter 3, they are not necessary for general serial data transmission. They could be added if one or two additional control signal paths would be use-

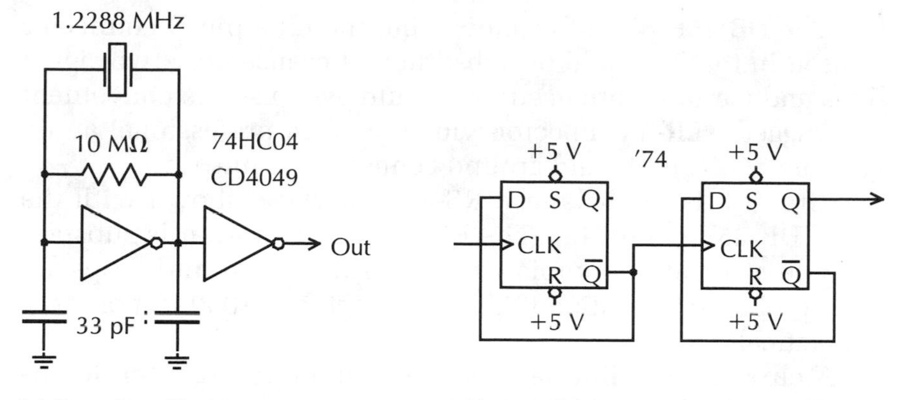

(a) Crystal oscillator (b) Divider for 4.9152-MHz oscillator

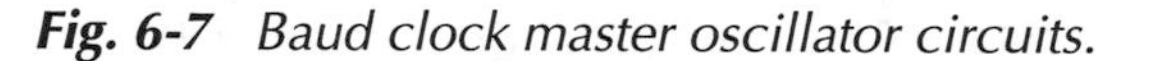

Fig. 6-7 *Baud clock master oscillator circuits.*

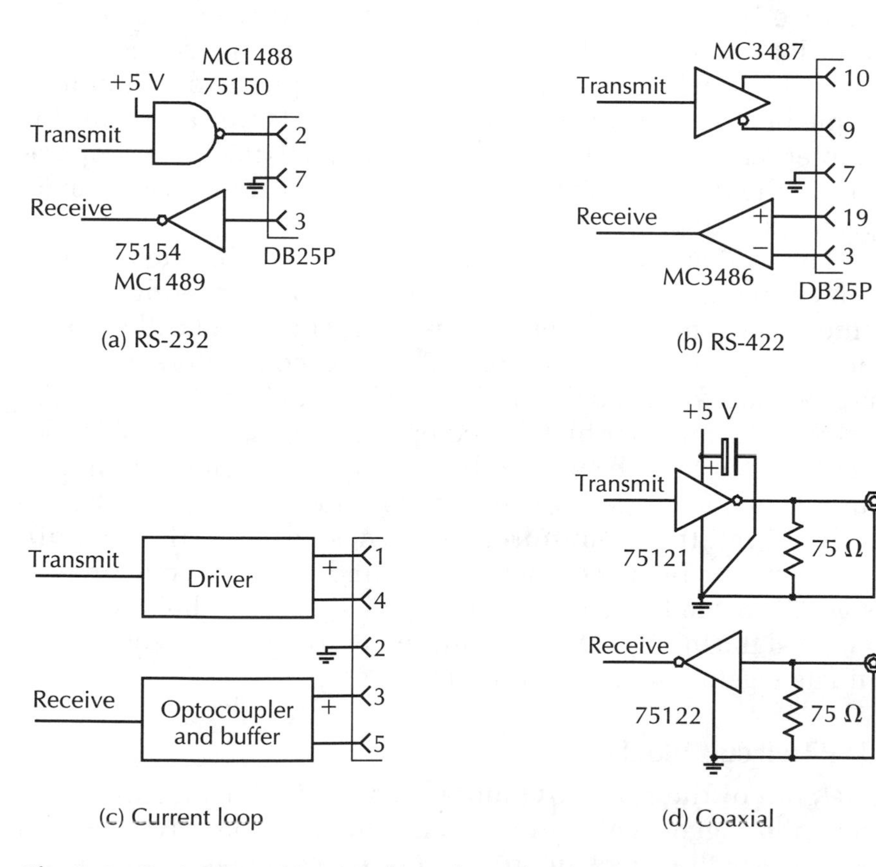

(a) RS-232 (b) RS-422

(c) Current loop (d) Coaxial

Fig. 6-8 *Serial line driver and receiver circuits: (a) RS-232, (b) RS-422, (c) current loop, (d) coaxial.*

ful. The DB-25P pinout is shown, but the nine-pin D connector as used in the AT could be substituted. Because only two signal lines and a ground are used, one could even use the convenient and rugged XLR-3 connector widely used in professional audio systems, with shield and ground connected to pin 1.

Circuit B is the balanced RS-422 interface, shown with the usual DB-25P connector. The RS-422 configuration is substantially less sensitive to noise than RS-232 and ground loops are not a problem. Operation with cable lengths of 1000 feet or more is practical.

A current loop interface is shown in circuit C. Simple circuits for a high-performance 20-mA current loop are given in chapter 7. There is no standard or customary connector, but the five-pin DIN (Hirschmann) connector could be used, as in the popular MIDI system. The current loop merits serious consideration. It is simple and inexpensive, has wide bandwidth, tends to produce little EMI, and is relatively insensitive to noise. The complete ground isolation provided by the optical couplers might be a primary concern in some applications, such as biomedical sensors. The CMOS logic circuits presented in this chapter feature low power consumption, making battery operation of the data input system practical in order to secure even greater isolation.

The standard coaxial interface is in circuit D. The termination shown is for 75-Ω cable such as RG-59/U or video cable, which seems more convenient than the 93-Ω cable called for in the IBM 360 I/O specification. The BNC connectors are commonly used for coaxial data links. The coaxial interface is the method of choice for high-speed operation (tens of megabits per second), but it can be used at lower rates also. Some care in layout is called for, especially for the transmitter. Because of the relatively high transient currents in these drivers the Vcc pin should have a bypass capacitor to ground (a 0.1-μF ceramic type is good) located as close to the driver IC as possible. So far as practical the input, output shield, and power grounds should terminate together at the IC ground pin.

UART circuit block

The core of the serial I/O module is the UART circuit block, shown in Fig. 6-9. A baud rate generator and line drivers would be added to this block in order to make a complete serial I/O interface module.

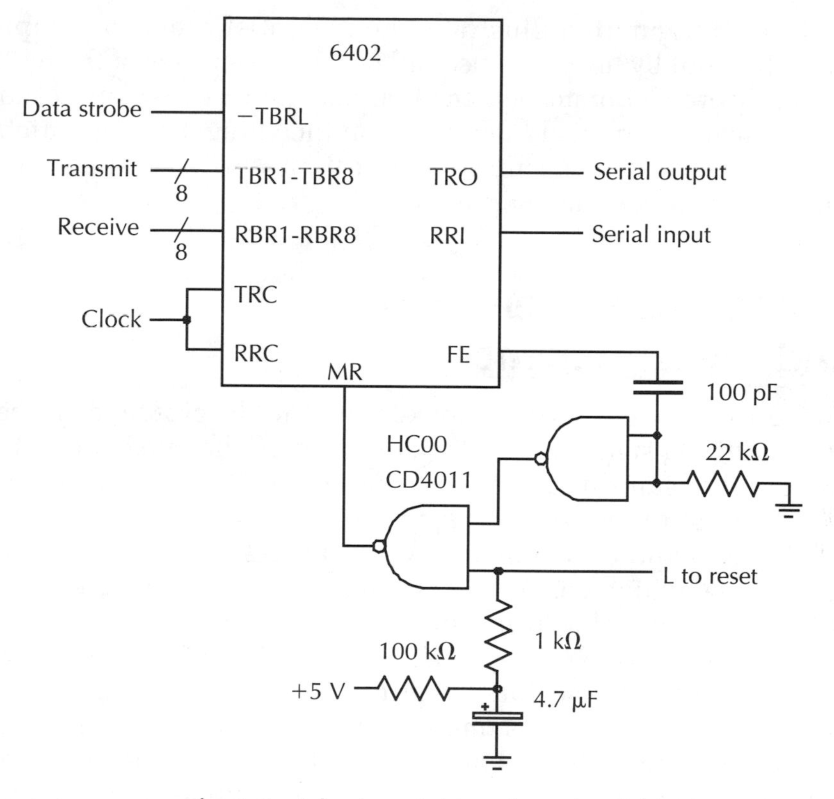

Fig. 6-9 *A basic serial interface circuit block.*

A negative-going data strobe pulse loads a byte to be transmitted into the UART. This is all that is necessary when the data rate is slower than the transmission. Otherwise either the TBRE or TRE could be used as a signal that thc UART is ready for the next byte.

The receiver connections shown in the drawing operate the receiver as a static data buffer: the data lines always contain the last byte that was received. This is suited to applications in which the receiver channel is used for status and control information. DR is connected to –DRR through an inverter so that each byte is acknowledged as it is received. The three-state receiver data lines are kept enabled by grounding RRD. If the receive data output is bussed, RRD would be used to control bus access.

For proper operation the 6402 must be reset after power-up. A simple delay circuit forces MR high for a short time when power is first applied. The framing error output is also connected to the reset circuit allowing a remote device to reset the UART by sending break.

The configuration illustrated here is satisfactory for simple serial I/O, but by no means uses all of the capabilities of the 6402 UART. More elaborate systems that use the receive channel for data transmission would use DR as an incoming data flag, might use TBRE to pace transmission, and the UART would perhaps share a data bus with other devices.

Digital instrumentation and measurement

The topics discussed in this book form a middle ground or go-between uniting signal and data acquisition on the one hand, with data analysis and reduction on the other. Our bailiwick was sketched out in Fig. 1-1 in chapter 1.

Signal acquisition covers two broad areas. There is *signal processing*, which consists of getting or transforming electrical signals into a suitable form; and there is *sensing* or *transduction*, which converts such nonelectrical quantities as force, pressure, temperature, and time, into electrical signals. Very often the output of a sensor requires some signal processing—scaling, filtering, and the like. There is an enormous amount of literature on these matters; two excellent introductory treatments are Sheingold (1974, 1980).

Data analysis is the oldest computer application. A truly vast array of techniques and algorithms have been devised to manipulate data in almost every way. The hardware modules and interface software examples in this book have been designed expressly to get information into a PC simply and efficiently, and to fit together well with analytical routines. A simple example of a data acquisition program is presented in chapter 7.

Continuous and discrete quantities

To reap the full benefit of data acquisition techniques, those in this book and commercial systems alike, some understanding of a few theoretical principles is necessary.

Sometimes the signal to be acquired is in digital form already, as in the case of an event totalizer or frequency counter. More often, though, the signal is an analog quantity and must be converted to digital form—digitized—in order to be manipulated in a computer. Strictly speaking, digital data are also "analogs"—they are numerical analogs or models or representations of something. For that matter, a digital data stream itself is really an analog electrical signal whose pattern is given a digital interpretation. (Good hardware designers know only too well that digital circuitry is just a particular application of analog circuitry as they wrestle with line ringing, thresholds, switching times, propagation delays, and the like.) It is better to contrast continuous with discrete information rather than analog with digital.

In a *continuous* system, some quantity (electrical, mechanical, or whatever) can support arbitrarily tiny variations in either its magnitude or its course in time. The signals in a *discrete* system, in contrast, can assume only certain values and no others, and there is always some smallest possible value (quantum). The smallest variation that can be represented in a discrete system is the size of its quantum. In the limit, the two converge: a continuous system is like a discrete system with an arbitrarily large number of quantization steps and arbitrarily fast sampling rate (large, but not infinite. Electrical interaction involves a vast but finite number of electrons or photons, and though the system bandwidth might be very great it is never infinite).

Digitizing

A discrete system can quantize magnitude, time, or both. An example of a system that quantizes only magnitude is a "flash" A/D converter, or even a simple voltage comparator that quantizes an input into two states. Systems that quantize time are called *sampled* systems. An example of a system that quantizes only time is FM stereo radio, which in essence samples the left and right audio channels alternately at a 76-kHz rate. A fully digital system quantizes both magnitude and time and, furthermore, usually represents the magnitude as a numerical value.

The quantization of the magnitude of a continuous signal necessarily throws away some information. The result is an approximation to the original. Because the tiniest changes cannot be represented exactly, quantization also introduces a small amount of noise (*quantization noise*). The ultimate resolution of a digital system is set by the number of bits used; N bits gives 2^N levels. The relative size of the quantum of a digital system is $1/2^N$. These things should be borne in mind when choosing any digital system.

Sampling—time quantization—likewise yields only an approximation to the original, but here the effects are more complicated. An important theoretical result, attributed to H. Nyquist, is that any signal can be sampled without loss provided that the sampling rate is greater than twice the highest frequency component in the signal. This minimum sampling rate is known as the *Nyquist rate*. If the sampling frequency is less than the Nyquist rate, the signal will be irreversibly corrupted due to frequency *aliasing*. An example of this is shown in Fig. 6-10. Note that the result, the lower curve, is still a sine wave—but at a lower frequency! It must be emphasized that Nyquist's criterion applies to the total frequency spectrum of a signal, including any noise in the signal, not just its basic repetition rate. Any frequency components in the signal that are above the Nyquist frequency will be "folded down" and mixed into the signal spectrum. For this reason, it is important in most applications to precede the A/D converter with a low-pass filter to suppress all frequencies above the Nyquist limit.

A more subtle corruption of a signal during A/D conversion is *aperture error*. A conversion takes a certain amount of time. If the signal is changing during this time, the final result will be incorrect.

To get a feel for the magnitude of the problem, let's look at some numbers. Let δ be the normalized smallest input increment that a system will process. For a digital system this is the resolution as a fraction of full scale. For a 4½-digit voltmeter, for example, full scale is 40,000 (that is, ±20,000), so $\delta = 2.5 \times 10^{-5}$. For an N-bit binary system, $\delta = 1/2^N$. An eight-bit system would have δ

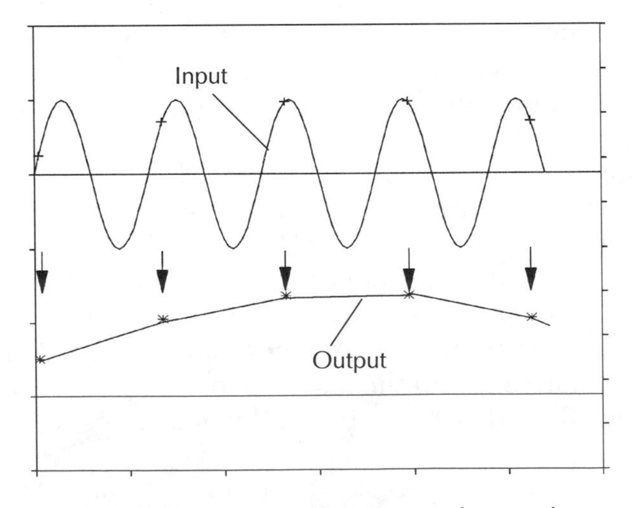

Fig. 6-10 *Illustration of frequency aliasing due to inadequate sampling rate.*

$= 1/2^8 = 1/256 \approx 0.0039$. To avoid error, the slewing rate (rate of change) of the input must be small enough so that it changes less than δ during the aperture time t_a (or, putting it the other way around, t_a must be short enough for a given slewing rate). In short, $t_a SR \leq \delta$ must hold.

If a sinusoidal input signal of frequency f and a peak amplitude equal to the full-scale value of the system is assumed, you can calculate the maximum allowable aperture time t_a as $t_a = \delta/2\pi f$. For an eight-bit converter digitizing a 100-Hz input the maximum t_a is 6.2 µs. This is not bad; a reasonably fast A/D converter can handle it. But now consider a sixteen-bit conversion of a 15-kHz input, as in a digital audio system. In this case the maximum t_a is 162 *pico*seconds—an impossible requirement.

You can also solve for the highest frequency f that a given resolution and aperture time will permit: $f = \delta/2\pi t_a$. Let's apply this to some of the hardware examples presented later in this chapter. One is an eight-bit A/D converter with a typical conversion time of 120 µs, so in this case f is 5.2 Hz. Another example is a 4½-digit voltmeter with an effective conversion (signal integration) time of 83.3 ms. The maximum f follows as 4.8×10^{-5} Hz, a period of 5.8 hours.

These dismal facts can be gotten around by using a *track-and-hold* circuit, which holds the value existing at the start of conversion in a capacitor, or the similar sample-and-hold circuit, which gets the value of the input just before each conversion. A very simple track-and-hold can provide an effective aperture time of a few microseconds; high-performance units achieve times in the tens of nanoseconds. The acquisition time—the time needed for the amplifiers and storage capacitor to reach the input value when switching to hold mode—must of course be shorter than the time between conversions.

The situation is more favorable when some form of integrating A/D converter is used. In the dual-slope converter widely used in digital voltmeters, the input remains active during conversion and changes in the input make some contribution to the result. The sigma-delta integrating converter requires no track-and-hold at all.

Signal reconstruction

For the sake of symmetry, D/A conversion should probably have been called "analogizing," but "reconstruction" has become the accepted term. The simplest method feeds each new data value directly to a D/A converter; the converter output changes immediately to the new value. This proceeds at the data rate, F_S. The output of the D/A converter is a step or staircase approximation to the original signal, as illustrated in Fig. 6-11. This sort of response is commonly known as the "simple hold" or "zero-order hold" operator.

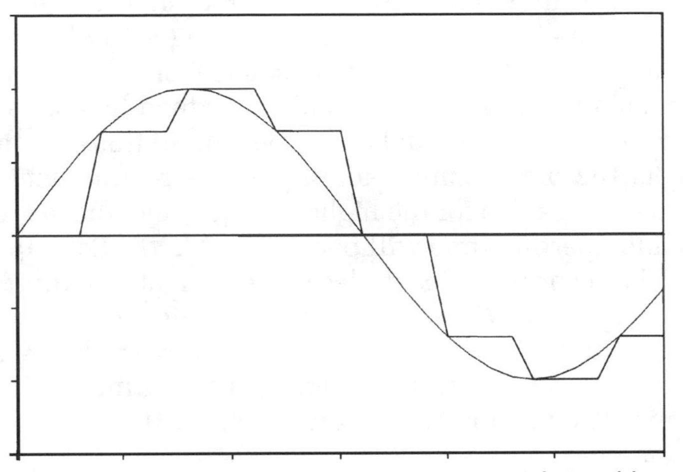

Fig. 6-11 *Staircase approximation to original signal because of sample-and-hold.*

At low signal frequencies the D/A converter output consists of many small steps that approximate the original signal fairly closely. At higher frequencies the approximation is rather coarse, as in Fig. 6-11 where the sampling rate F_s is eight times the signal frequency f_o. The frequency spectrum of this step waveform is shown in Fig. 6-12, normalized such that the Nyquist limit ($F_s/2$) is 1 Hz. Sure enough, the main component is the original signal frequency f_o at 0.25 Hz, but there are significant components at higher frequencies. Compare this to Fig. 6-13 which is the spectrum of a reconstructed signal at $f_o = 0.125$ Hz ($F_S = 16f_o$); the beneficial effect of the greater number of smaller steps is evident.

Sending the output of such a D/A converter through a low-pass filter will suppress the high-frequency garbage and smooth out the steps in the reconstructed signal waveform. The characteristics required of such a filter depend on how close to the Nyquist limit you are working. The spectrum for a signal at 0.75 of the Nyquist limit ($F_S = 2.66f_o$) is shown in Fig. 6-14. Compare it to the previous two spectra and note that as the signal fre-

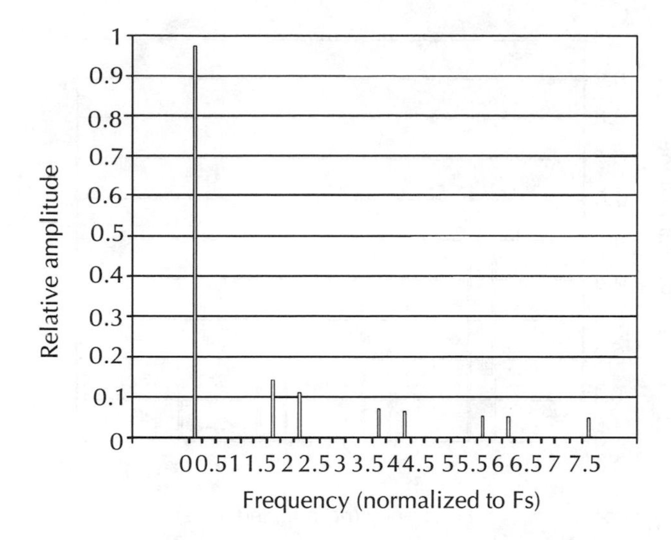

Fig. 6-12 *Spectrum of 8× sampling.*

quency increases toward the Nyquist limit the harmonics move down toward the Nyquist limit and also increase in magnitude. Systems that must operate close to the Nyquist limit must use sharp cutoff filters to achieve a clean output signal. The trade-offs are like those in antialiasing filtering but less severe; here, the principal effect of too little filtering is high-frequency artifacts in the output.

But life isn't quite that simple. A second thing to notice in comparing the spectra in Figs. 6-12 to 6-14 is that the magnitude

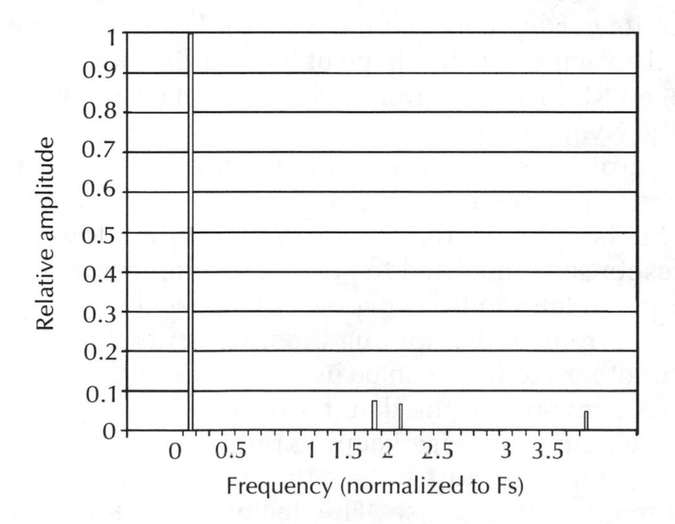

Fig. 6-13 *Spectrum of 16× sampling.*

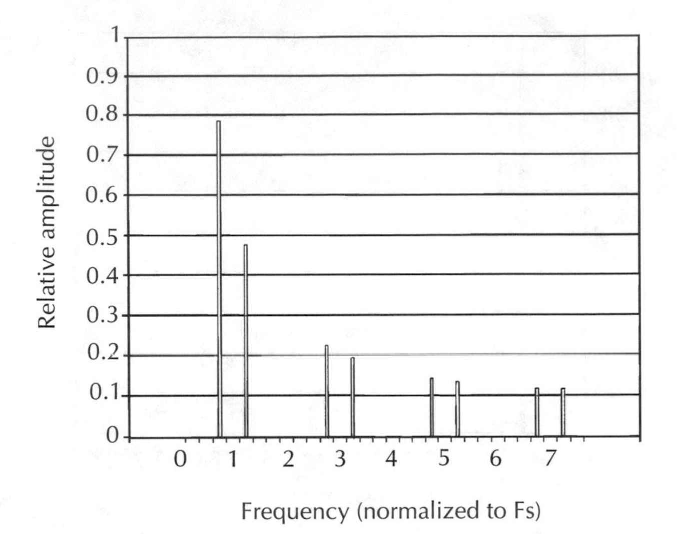

Fig. 6-14 *Spectrum of 2.66× sampling.*

of the original signal component at f_0 decreases somewhat as f_0 approaches the Nyquist limit. This mild low-pass effect is due to the holding of the output between samples. The output from the D/A converter can assume a new value only every T_S seconds, where $T_S = 1/F_S$. Clearly the output can contain no signal change shorter than T_S (the aperture effect in reverse, so to speak). Specifically, the response as a function of signal frequency f of the simple hold operator is proportional to $\sin(\pi(f/F_S)) / \pi(f/F_S)$. At the Nyquist frequency, where $f/F_S = 0.5$, the original signal is 0.6366 of its correct amplitude (–3.9 dB). For frequencies up to the Nyquist frequency, the shape of the $\sin(x)/x$ response is close to that of an RC low-pass circuit with a –3-dB cutoff frequency of 0.83 of the Nyquist frequency.

For simple systems, these effects might be unimportant. Where better performance is required, there are two avenues of attack. One is to follow the D/A converter with a low-pass filter whose response is modified to give flat overall response. An example is given later in this chapter. A more accurate and elegant method is to reduce the holding time, which is the real culprit. Latches or other methods can be used to connect the digital input of the D/A converter to the data for only a short time after each new data value, otherwise keeping the D/A input at zero. The D/A output will be short pulses that give much better reconstructed frequency response. This technique is known as *oversampling* and is the method of choice where performance is critical. An even more powerful technique combines oversampling with a digital interpolating filter prior to the D/A conversion. Superb frequency and phase characteristics can be obtained this way, as in the better audio CD players. A good introductory treatment of these more powerful digital techniques is in Proakis (1988).

Some analog filters

The importance of signal filtering in digital data acquisition systems should be clear. The filters used can have a pronounced effect on the overall system, and some care is needed in selecting them. This subject tends to be glossed over to some extent in many data acquisition handbooks, so it is worthwhile to give it a closer look.

Let's take a moment, first of all, to consider filtering in a very general way. Filtering is by definition modifying the frequency spectrum of a signal. If you happen to have the spectrum of a signal available you could simply adjust the individual frequency components to get the result you want. This is rarely the case, of course. Spectra do not strictly exist as physical quantities. A spectrum is a function or set of values in the frequency domain, which is actually a mathematical construct (and a very useful one). All real-world signals, whether analog or digital, are a function of time; their actual existence is in the time domain.

You can, however, obtain representations of the spectrum of a signal by using a spectrum analyzer for analog signals and the Fourier transform for digital signals. You could therefore filter a signal by obtaining its spectrum, manipulating the frequency components, and converting the result back to the time domain. If the signal is in digital form you could use the forward and inverse Fourier transforms to move between the time and frequency domains. An analog system could, at least in principle, simulate this by means of a spectrum analyzer controlling a bank of oscillators.

You can work directly on the signal itself in the time domain by *convolving* the signal with a filtering function. In a digital system, the filtering function is held in an array of coefficients. Each output value is generated by taking each input value and looping through the coefficient array, multiplying the input value by each coefficient and adding it to a running sum; the sum becomes the output value. (The "moving average" method of data smoothing is a crude application of this technique.) In an analog system, a tapped delay line is used to get successive input values that are then summed; the values of the input resistors that are of summed correspond to the filter coefficients. Analog filters of this sort—transversal filters—have been widely used in video and radar systems since the 1940s.

The accuracy of convolutional filters depends on the number of coefficients or *taps*. Practical considerations limit analog transversal filters to perhaps a dozen or so taps. The main constraint in the digital version is the time required to do the considerable number crunching. Specialized microprocessors —digital signal processors (DSPs)—can do this at high speed. Now that inexpensive DSP chips are readily available you can expect a rapid expansion in the use of this powerful technique.

Convolutional filters apply a filtering function directly to a signal. A different and somewhat indirect method is to use storage to average a signal continually over a short time interval. This is what the familiar RC and LC analog filters do, using capacitors and inductors as short-term storage elements. The digital equivalent, the recursive filter, stores a running value in a temporary variable. It requires very few computations.

The output of a storage filter decays toward zero after the input is removed, but (theoretically) never quite reaches zero. It has *infinite impulse response* (IIR). The output of a convolutional filter, on the other hand, is indeed zero once all the values currently in the pipeline (shift register or delay line) are processed. It has *finite impulse response* (FIR).

An analog RC or LC filter is simpler, and a recursive digital filter requires much less number crunching than their convolutional counterparts. The price that is paid for this simplicity and economy is a much more limited and rather inflexible range of filtering characteristics. In particular, the magnitude and phase response cannot be chosen independently. Furthermore, the phase response is not linear. Lagging phase shift in the frequency domain corresponds to delay in the time domain. A constant delay for all frequencies corresponds to a lagging phase shift directly proportional to frequency. The rate of change of phase θ in radians with frequency f in Hz is the *group delay* t_g in seconds: $t_g = -\, d\theta/d(2\pi f)$. If the group delay is not constant, different frequencies will pass through the filter at different rates, further distorting the shape of the waveform. Storage filters are not phase linear and thus do not have constant group delay. Convolutional filters, on the other hand, can be designed to have either constant or varying group delay.

A filter necessarily changes the shape of the signals passing through it. The kind of change depends on the rate of change of the filter response with frequency. In the case of low-pass filters, a gradually decreasing response will smooth the waveform and reduce noise, but will also attenuate frequencies for some distance below the cutoff frequency. A sharp cutoff response, on the other hand, has nearly uniform transmission up to the cutoff frequency, but there will be overshoots or "ringing" in the output waveform. This is not because of any resonant circuits in the filter but due to the rapid change in the filter response. (It occurs with both storage and convolutional filter types.) There is an unavoidable trade-off between sharpness of cutoff and waveform distortion. The exact relationship is complicated, but as a rule of thumb, a filter with a response near cutoff that is significantly sharper than that of a simple RC circuit will have some overshoot.

Low-pass filters

Filter theory is a rich and challenging specialty. Here I can only touch on a few simple results that are directly useful in practical digital data acquisition systems. The filters described here are all-pole low-pass analog filters. Such filters are characterized by a cutoff frequency F_C above which the transmission drops toward zero. The attenuation above the cutoff frequency is a function of the *order* or number of response poles.

A brief explanation of a few main points might be helpful for those who are unfamiliar with complex frequency analysis. *Response poles* are frequencies at which the slope of the response curve changes, increasing by 6 dB per octave and thus bending more downward on a frequency response graph. Poles can be either single real poles or complex pole pairs. The simple RC low-pass has a single pole. It is 3 dB down at its pole frequency, and at frequencies above this its response falls at 6 dB per octave. Pole pairs, on the other hand, are like resonant circuits. The response falls at 12 dB per octave above the pole frequency. The behavior at the pole frequency depends on the damping factor α. For $\alpha = 2$ the response is the same as two identical single-pole circuits. For $\alpha < 1.552$ there is a peak at the pole frequency and the height of the peak increases as α decreases. (There are response zeroes as well as poles, but these filters do not have any. A real zero decreases the slope by 6 dB per octave and a zero pair creates a response notch.)

Filters are formed from combinations of single poles (real poles) and pole pairs (complex poles). A wide variety of overall response curves are possible by selecting various pole frequencies and damping factors. The cutoff frequency is commonly defined as the frequency at which the overall response is 3 dB down. The criteria for selecting poles and damping are called the filter alignment. Different alignments give different magnitude and delay characteristics. Quite a few alignments have been developed but only two are considered here.

One of the oldest and most widely used is the maximally- flat magnitude alignment developed by Butterworth in 1930. It is an exact answer to the question, "How do we get the flattest possible passband response?" It has flat response nearly up to the cutoff frequency and fairly rapid attenuation thereafter. For many applications Butterworth filters are a good compromise between sharpness of cutoff and linearity of group delay. Butterworth filters have moderate overshoot.

A different approach is used in the maximally-flat delay alignment by Thomson. This alignment, also known as the Bessel alignment, maximizes the flatness of group delay at the expense of a rather gradual slope near cutoff—about the same as a single-pole RC circuit. Its lack of a sharp cutoff makes it a poor choice for such uses as antialiasing. Bessel filters have no overshoot.

It is possible to make a high-order filter from identical single poles only. This is known as the coincident pole or synchronously tuned alignment. Up to the cutoff frequency its frequency response is close to the Bessel alignment, but its group delay is not as constant. Above cutoff its slope is more gradual than either Bessel or Butterworth.

Figure 6-15 compares the response curves of several different alignments and orders: a single pole (1P), two coincident or synchronous poles (2C), a second-order Butterworth (2B), and finally a fifth-order Butterworth (5B) filter. The curves are shown with a linear frequency scale to reveal more detail near the cutoff frequency. The Bessel alignment is close to the 2C curve for all orders.

Practical filter circuits

There are many filter circuits. The ones discussed here are among the best. They are also in a form that allows a simple design procedure to be used. They were selected on the basis of the following criteria:

- The filters are canonical; that is, they use the minimum possible number of RC time constants (one per pole).
- All of the time constants in a section are equal. There is no problem of "element spread."
- The value of the frequency-determining capacitors is a free variable. This is helpful because there are fewer standard values for capacitors.
- Resistor ratios are made unity (matched pairs) wherever possible.

Table 6-3 Normalized filter parameters.

Order (poles)	Butterworth		Bessel	
	Fp	α	Fp	α
2	1.00000	1.4142	1.2742	1.7320
3	1.0	R	1.3248	R
	1.0	1.0	1.4499	1.4471
4	1.0	1.8478	1.4324	1.9160
	1.0	0.76540	1.6059	1.2424
5	1.0	R	1.5047	R
	1.0	1.6180	1.5588	1.7745
	1.0	0.61800	1.7581	1.0911
6	1.0	1.9319	1.6065	1.9596
	1.0	1.4142	1.6919	1.6361
	1.0	0.51719	1.9078	0.97722
7	1.0	R	1.6871	R
	1.0	1.8019	1.7191	1.8785
	1.0	1.2470	1.8254	1.5132
	1.0	0.4450	2.0528	0.88786

Table 6-3 shows the normalized pole frequencies f_p and damping factors α for Butterworth and Bessel filters of orders 2 through 7. An entry of R for damping indicates a single (real) pole. Frequencies are normalized to give an overall –3-dB filter cutoff frequency f_c of 1 Hz. To scale to a different frequency, multiply the frequency values given by the desired f_c. Two trivial alignments are not shown. A first-order filter is just a single pole at f_c. A coincident pole (synchronous) filter of order N is a cascade of N single-pole circuits, each with a pole frequency of $f_c/[\sqrt{(2^{1/N}-1)}]^{1/2}$.

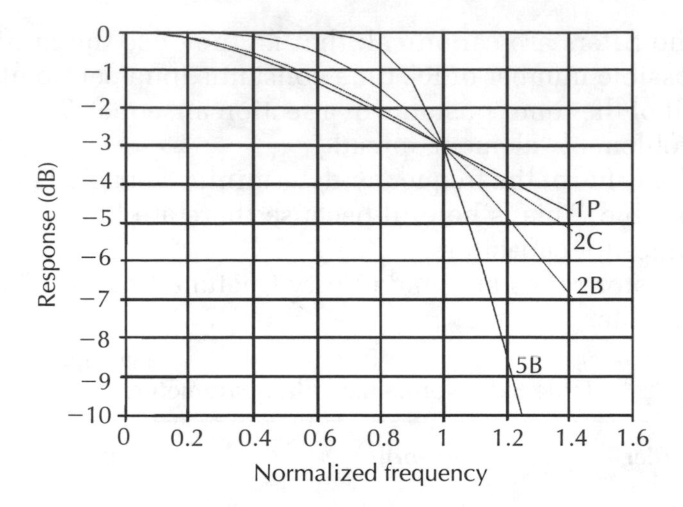

Fig. 6-15 *Filter frequency response curves. 1P, single-pole; 2C, coincident-pole; 2B, second-order Butterworth; 5B, fifth-order Butterworth.*

The design procedure is straightforward, not unlike working from a cookbook. Filters of order 2 and higher consist of pole-pair circuits strung together in cascade, with an additional single pole for odd-order filters. A fifth-order filter, for example, is made up of a single pole followed by two pole-pair blocks. Each section has a particular pole frequency f_p (not necessarily the same as the overall filter cutoff frequency) and damping factor α. These values are listed in Table 6-3. The sections can be connected in any order but for greatest dynamic range the sections should follow one another in order of decreasing damping factor. In an odd-order filter the single real pole should be placed first.

A complete filter is designed section by section as follows:

1. Get the normalized parameters for the filter alignment you want to use from Table 6-3. Multiply the normalized frequencies given in the table by the required filter cutoff frequency f_c. This gives the actual pole frequency f_p for each section.
2. Each filter section is then designed using the component value equations given in the description of each circuit configuration. First, choose a convenient value for the frequency-determining capacitor Cf, and then calculate the corresponding value of Rf. Next, calculate the remaining component values, all of which are resistors. The value of

all resistors should be kept between 10 kΩ and 1 MΩ for FET amplifiers, or 10 to 100 kΩ for bipolar amplifiers. The value of Cf should be at least 470 pF to minimize inaccuracies due to stray and amplifier capacitances. Inexpensive ICs such as the TL070 and LF347-353 families work well up to 30 kHz or so, above which amplifier phase shift begins to degrade the accuracy.

Single poles Two circuits for realizing a single pole are shown in Fig. 6-16. One is a simple RC circuit with a unity-gain buffer for isolation. The circuit must be driven by a low-impedance source such as an op amp. The other uses an inverting amplifier that provides greater isolation and allows the gain H_0 to be other than unity. In both circuits, the pole frequency $f_p = 1/(2\pi RfCf)$. The design procedure is as follows:

1. Choose a value for Cf.
2. $Rf = 1/(2\pi f_c Cf)$
3. For the inverting circuit, choose the gain H_0. Then, $H_0 = Rf/Ri$ and $Ri = Rf/H_0$.

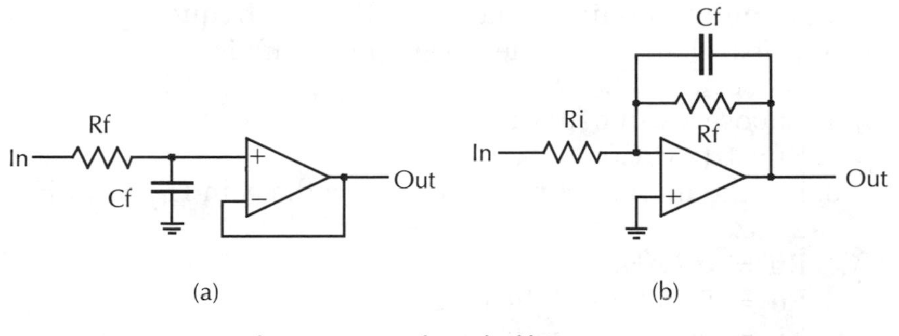

Fig. 6-16 *Single-pole filter circuits.*

Pole pairs The "state variable" configuration shown in Fig. 6-17 is used. The state variable circuit has many advantages over other active filter configurations. It is very stable, it depends chiefly on resistor ratios, the frequency and damping adjustments do not interact, the overall gain can (within limits) be freely chosen, and the frequency-determining RC combinations are equal regardless of damping. It realizes low-pass, band-pass, and high-pass responses simultaneously. The circuit requires three amplifiers, but this is not a serious drawback in these days of inexpensive quad IC op amps. Many analyses of the state vari-

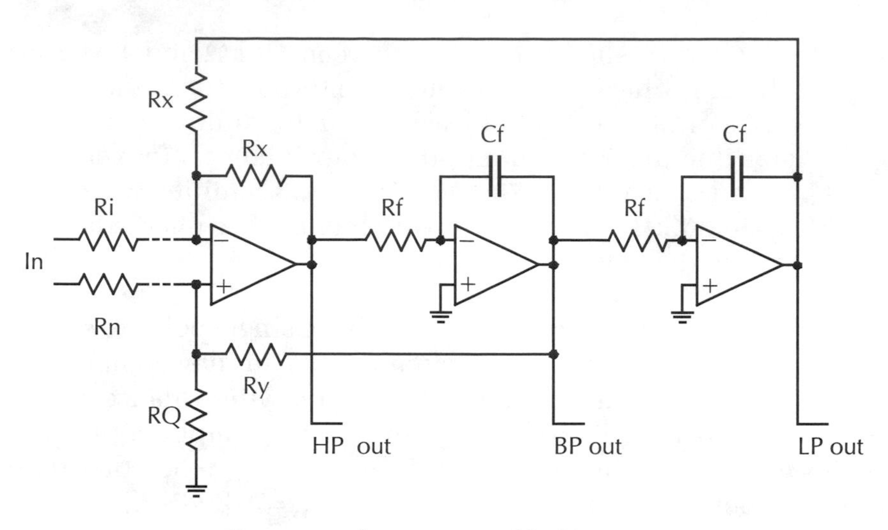

Fig. 6-17 *The state-variable filter circuit.*

able circuit have been published; there is a concise treatment in Graeme (1971).

The pole frequency $f_p = 1/(2\pi RfCf)$. The circuit can be used in either an inverting or a noninverting configuration. There is a restriction on the noninverting form: the low-frequency gain H_0 must be less than $2 - \alpha$. The design procedure is as follows:

1. Choose a value for Cf.
2. $Rf = 1/(2\pi f_c Cf)$
3. For the noninverting form, choose the gain H_0, with $H_0 \leq 2 - \alpha$.

   $Rn = \alpha Ry/H_0$
   $Rq = Ry/\{ [(2 - H_0)/\alpha] - 1 \}$
   Ri = not used, omit

   For the inverting form, choose the gain H_0.

   $Ri = Ry/H_0$
   $Rq = Ry/\{ [2 + H_0)/\alpha] - 1 \}$
   Rn = not used, omit

A phase compensated filter

In ordinary filters of the kind considered here, the magnitude and the phase responses are uniquely related to one another. As the response magnitude improves (that is, the sharpness of the cutoff increases) the group delay deteriorates, and vice versa. Although this problem could be sidestepped by using a transversal filter, it is often desirable or necessary to stick with the simplicity of the conventional low-pass structure. A compromise is to add a phase compensation network to a conventional filter. This does not eliminate the trade-off between the magnitude and delay responses, but it can improve it quite a bit.

The design of phase compensation is rather complicated; there is no general analytic solution. But because a constant theme of this book is getting high performance out of simple and inexpensive systems, it seems appropriate to offer a useful working example. The circuit of a third-order delay-corrected low-pass filter building block is shown in Fig. 6-18. The magnitude of the frequency response, shown in Fig. 6-19, is that of a third-order Butterworth filter. It provides good pass-band response with 18 dB/octave attenuation above cutoff. The dashed curve is an expanded plot (1 dB per division) to show the behavior around cutoff more clearly. The flatness of the group delay is similar to a third-order Bessel filter and is shown in Fig. 6-20. The upper pair of curves are the group delay of this filter (solid curve) and the percent deviation from constant delay (dashed curve). The two lower curves are for a standard uncompensated third-order Butterworth filter for comparison. The group delay of the compensated filter is markedly flatter.

Component values for this filter are as follows. The –3-dB cutoff frequency f_C is $1/(2\pi \mathrm{RfCf})$. For a given value of Cf, Rf = $1/(2\pi f_C \mathrm{Rf})$. The actual values of the two resistors Rx is not critical—somewhere in the range of 10 to 22 kΩ is a good choice—but they should be matched as closely as possible; and likewise for the Ry pair. The low-frequency gain $H_0 = 2\mathrm{Rf}/\mathrm{Ri}$, so for unity gain Ri should be twice Rf. The low-frequency group delay is $t_g = 4/(2\pi f_C) = 4\mathrm{RfCf}$.

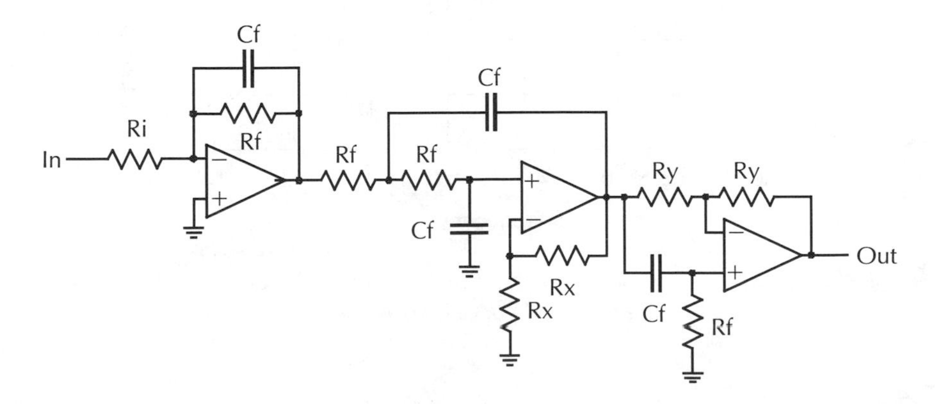

Fig. 6-18 *Third-order Butterworth filter with phase correction.*

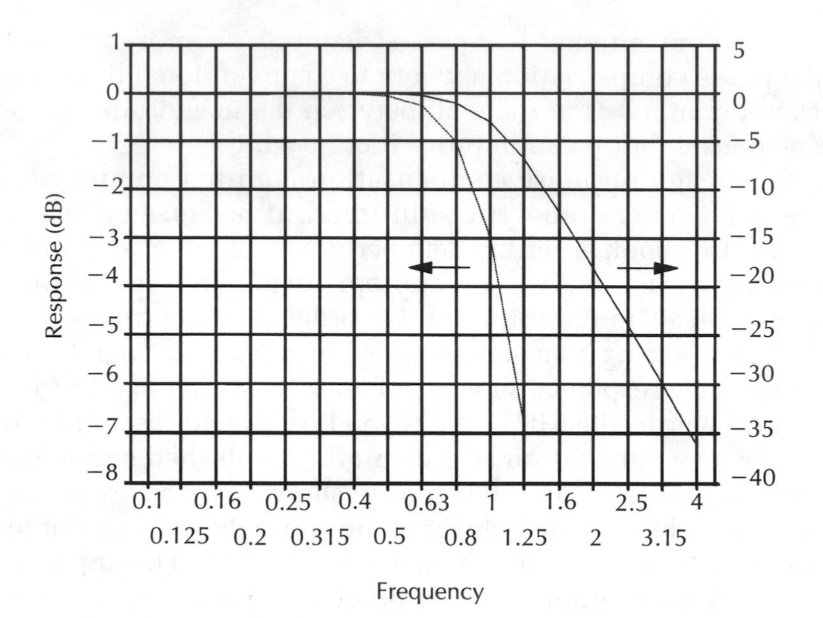

Fig. 6-19 *Frequency response of phase corrected third-order filter. Dotted line is for expanded scale on left.*

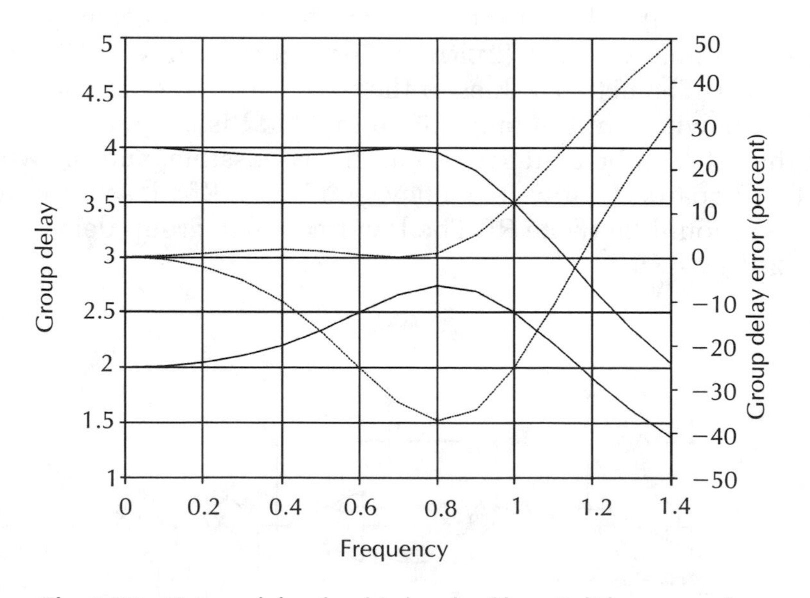

Fig. 6-20 *Group delay for third-order filter. Solid curves, phase corrected; dotted curves, conventional Butterworth.*

A/D converter circuits

This section presents a versatile general-purpose eight-bit A/D converter circuit block. For applications where eight-bit resolution is sufficient and sampling rates of 8 kHz or less are adequate, it offers excellent performance at low cost. Later in this chapter I examine a different A/D circuit block that offers very high resolution but very low sampling rate.

Principles of A/D conversion

There are many methods of A/D conversion in use, all with different strengths and weaknesses. Let's take a quick look at the major methods before turning to the A/D circuit block.

Direct conversion The simplest A/D converter, at least conceptually, consists of a group of comparators. The signal is applied to one input of each comparator, and the other comparator input is connected to a tap on a voltage divider. The comparators classify the amplitude of the signal; those whose threshold voltage is less than the input voltage are on, and those above are off. An array of logic gates converts the comparator outputs to a binary code. The output of such a converter is immediate, except for the small response time of the comparators and the switching time of the logic gates, and it is commonly called the *flash* converter. Its output is continuous and tracks the input signal.

Flash converters are widely used in video A/D converters because of their speed. A fatal (almost) disadvantage is that coding N bits requires $2^N - 1$ comparators. This is practical with monolithic circuits, at least up to a point; eight-bit devices (255 comparators) are available, but a sixteen-bit device (65,535 comparators!) seems a distant goal. Quite inexpensive flash converters are now available, such as the Harris (RCA) CA3306 six-bit CMOS converter which will run at sample rates as high as 12 MHz.

Integrating conversion Sometimes the long way 'round is the shortest way home. The integrating converter uses an indirect approach that achieves very high accuracy and resolution with relatively simple circuitry. Its main drawback is that it is much slower than other types.

A widely used form of integrating A/D converter is the dual-slope converter. It consists of an integrator, a comparator, electronic switches, a counter, and clock and control circuits as shown in Fig. 6-21. A conversion cycle consists of several steps:

1. Conversion starts at T0 with switch S1 connecting the input signal to the integrator. The integrator output increases linearly in proportion to the input voltage—a higher Vin produces a larger integrator output and thus a steeper slope.
2. After a preset number of clock counts, at time T1, switch S1 connects the integrator to a reference voltage. A counter is also enabled; it will hold the final result. The control circuits now monitor the comparator output. The reference Vref is a constant voltage opposite in polarity to Vin, so the integrator output decreases at a known and constant slope. During this time the counter is incrementing.
3. When the comparator detects that the integrator output has reached zero, at T2, the value in the counter is latched and the conversion is complete.

When the input voltage Vin is large, it takes longer for the integrator to reach zero during Td and the counter holds a large

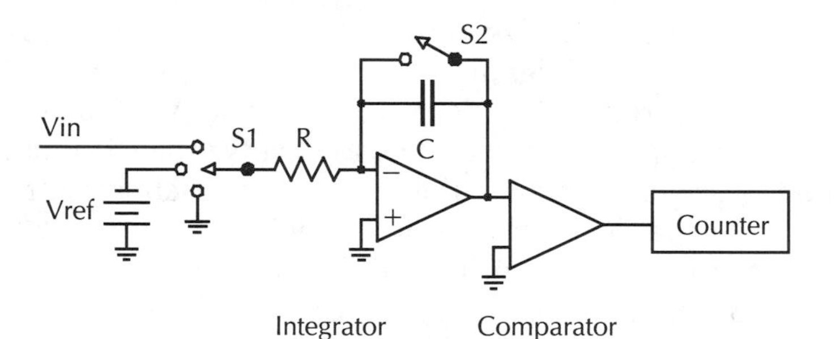

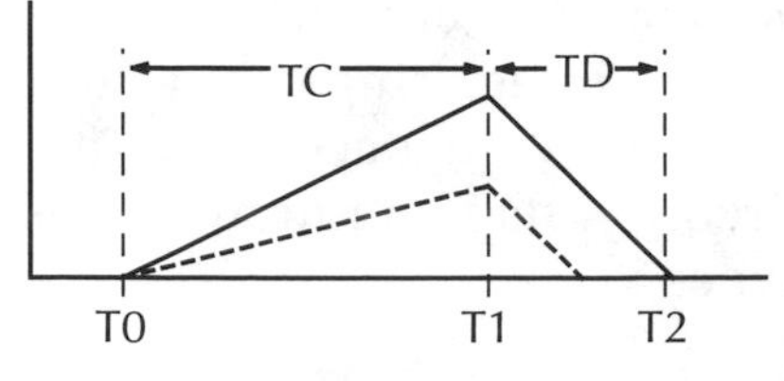

Fig. 6-21 *Block diagram of a dual-slope A/D converter.*

count. When Vin is small, the integrator output will rise only a little during Tc, as shown by the dotted lines in Fig. 6-21. The integrator quickly ramps down to zero during Td and the counter holds a small count. The count is in fact directly proportional to the ratio between Td and Tc. Because Tc is proportional to Vin and Td is proportional to Vref, you are implicitly finding the ratio Vin/Vref. A dual-slope converter is thus *ratiometric*. It measures the ratio of an unknown voltage and a known reference voltage.

The fact that the dual-slope converter operates primarily with ratios gives it some valuable properties. Its accuracy depends only on the linearity of the integrator, not its absolute time constant. Because many of the elements in the analog signal path are used for both Vin and Vref, their inaccuracies largely cancel out. As long as the clock frequency is constant during a conversion, the actual frequency of the clock does not affect the measurement. The output counter does not control the operation of the converter, so any kind of counter can be used—natural binary, 2s-complement binary, BCD, and so on.

Practical dual-slope converters usually incorporate some additional goodies, such as automatic polarity detection and a short autozero cycle between conversions. Autozero consists of a dummy conversion cycle with the input connected to ground during Tc. The resulting value is then subtracted from the actual measurement. The digital voltmeter described later uses a dual-slope converter with these features.

Servo converters Another widely used method is the servo converter illustrated in Fig. 6-22. A counter is connected to a D/A converter, which is connected to a comparator in a servo or feedback arrangement. At the beginning of a conversion, the counter is cleared and thereafter is incremented by a clock. The D/A output will increase until it is equal to the input voltage Vin. At this point the comparator switches, disconnecting the counter from the clock. The count in the counter is the digital value of the input. The method is simple and accurate, but inefficient. The conversion time is a function of the input level and is much longer for inputs near full scale.

A modification of the straight servo converter, the *successive-approximation converter*, overcomes these problems and is now the workhorse of the industry. Instead of counting up bit by bit, the converter first takes just the most significant bit and tests whether the input is above or below half-scale. That cuts the

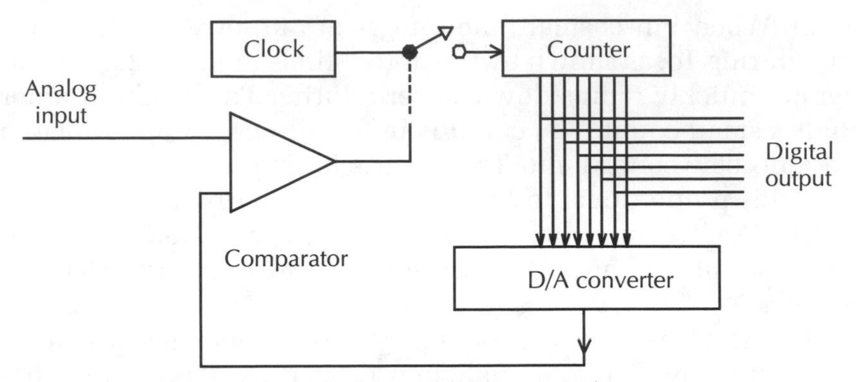

Fig. 6-22 Block diagram of a servo A/D converter.

number of clock counts needed in half. It then tries the next most significant bit, again reducing the number of clock counts needed by half, and so on. An *N*-bit conversion takes only *N* clocks, regardless of the input level—a vast improvement in efficiency. This is accomplished by replacing the binary counter of the basic servo converter with a successive approximation register (SAR), essentially a shift register with latches.

Sigma-delta conversion This converter (Fig. 6-23) is something of a cross between an integrating and a servo converter. An integrator and comparator are used, and there is feedback through a D/A converter. Note, however, that the D/A converter is a one-bit converter, and that its output is subtracted from the input. The feedback loop tries to keep the integrator output at zero, which occurs when the average D/A output equals the av-

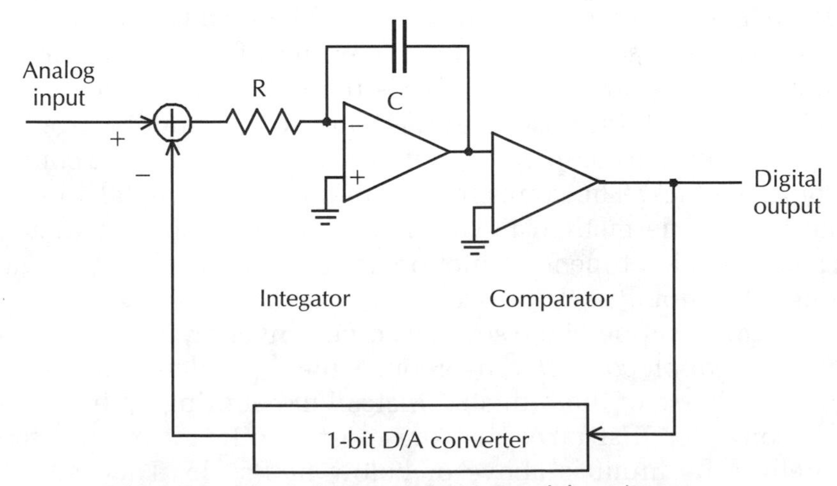

Fig. 6-23 Block diagram of a sigma-delta A/D converter.

erage input voltage. A one-bit A/D conversion might not seem very useful, but actually the sequence of 1s and 0s in the output stream contains a high-resolution representation of the input. An ordinary digital or analog output can be constructed from this stream by various methods including low-pass filtering. Because the one-bit conversion can run at very high sample rates, both high resolution and fast conversion are possible.

This method will become increasingly important. Sigma-delta converters and DSP chips are already inexpensive enough for use in realizing eighteen-bit conversions at a 44.1-kHz sample rate in consumer digital audio products.

An eight-bit A/D converter circuit block

Now for an actual circuit building block. This A/D converter block offers eight-bit conversion at sampling rates up to 8 kHz, allowing signals up to 4 kHz to be digitized. This is adequate for many instrumentation systems, and even telephone-grade audio. It can be used with the serial and parallel interface circuits described earlier in this chapter for sending data to a PC. It provides a basic 0- to +5-V analog input.

The block is built around the ADC080x, an eight-bit successive-approximation A/D converter using CMOS technology. This popular chip is inexpensive and easy to use and provides moderate speed (up to 8000 conversions per second). The device is available in four grades of accuracy. The ADC0802 provides ±½ lsb accuracy and is a good choice for general-purpose use. The ADC0801 offers ±¼ lsb accuracy, the ADC0803 ±¾ lsb, and the least expensive ADC0804 provides ±1 lsb.

Figure 6-24 shows the ADC0802 in a basic self-clocking circuit. The device is designed to facilitate operation with a microprocessor. The end-of-conversion ("data ready") signal is latched for ease of use as a processor interrupt and is brought out to the-INTR pin. The converter has three-state outputs to simplify connection to a bus; –RD and –CS must be low to enable the output lines. The –WR input acts like a start-conversion input. The analog signal inputs and reference trim are discussed later.

The device is optimized for a clock in the range of 600 to 640 kHz. A clock frequency less than 600 kHz increases the conversion time which, as noted earlier, should ordinarily be kept as short as possible to minimize aperture error. Above 640 kHz the conversion accuracy is somewhat reduced. There is a built-in RC oscillator circuit that runs at a frequency of about 1/RcCc; the values shown pro-

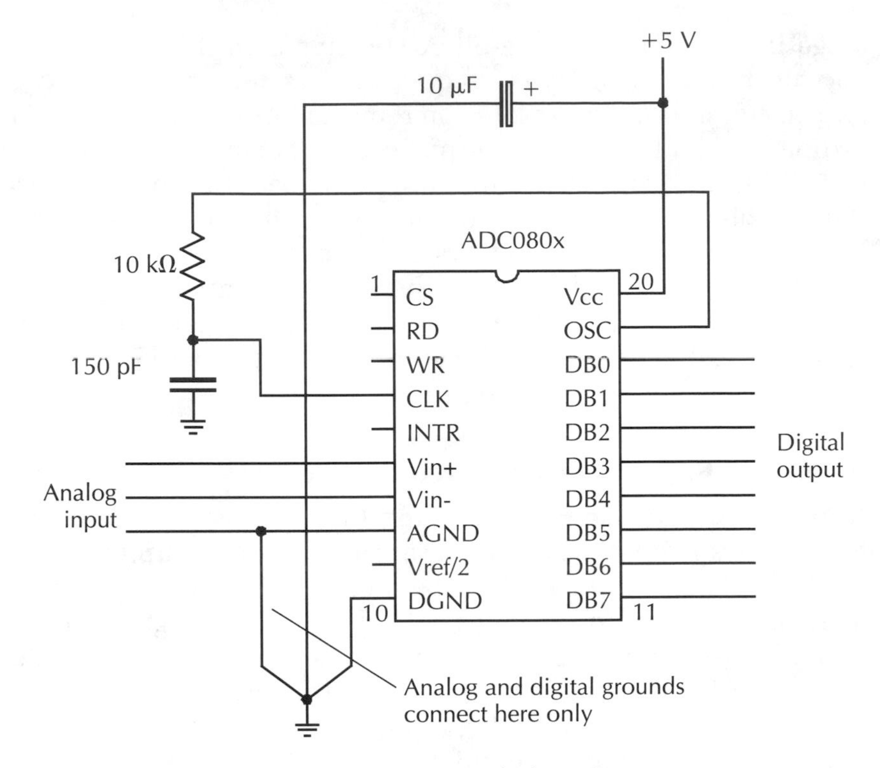

Fig. 6-24 *The ADC080x eight-bit A/D converter.*

duce approximately 600 kHz. An external TTL-level clock can be used instead, in which case it is connected to CLK, pin 4, and Rc and Cc are omitted. An example in which a separate clock would probably be used is a converter with serial data output using the 6402 UART presented earlier in this chapter. The 614,400 Hz available from the baud rate divider chain would do nicely as an ADC0802 clock, giving 114 μs average conversion time.

The basic converter block By adding some simple support circuits to the basic ADC0802 you get a very useful A/D converter block. Figure 6-25 shows the basic block. The main addition is a latch to control a track-and-hold circuit. The HOLD output is high, and the -HOLD output is low, during the conversion.

Conversions are initiated by an external start pulse. The converter is in a standby mode as long as START is low. The conversion cycle begins when START goes high. START must remain high during the conversion; if it goes low, the conversion in progress will be abandoned. When the conversion is complete, EOC (end-of-conversion) goes low for eight clocks. The device is ready for a new conversion as soon as EOC goes low. To reset for the

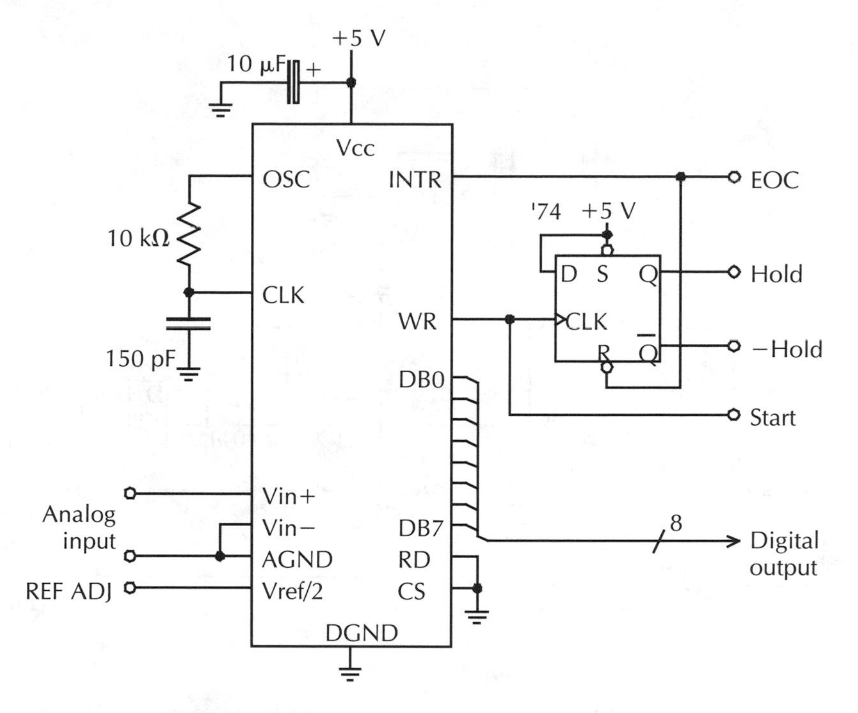

Fig. 6-25 *A basic A/D converter circuit block using an external start-conversion signal.*

next conversion, START must be low for at least 500 ns. (If START goes low while EOC is still low, EOC will reset to high.)

The actual conversion begins one to eight clocks after START goes high and runs at eight clocks per bit; a conversion takes sixty-six to seventy-three clocks. This slight jitter in conversion time can be avoided if an external clock is used and the START pulse is synchronized to it.

Figure 6-26 shows a very useful self-starting version of the basic block. The conversion rate is controlled by a 555 timer. A simple RC circuit generates a narrow start pulse (about 1.8 μs) after the 555 times out. Forcing the run input low suspends conversions and puts the converter in standby mode.

The timing is detailed in Fig. 6-27. The 555 delay time Td can be calculated for a desired average cycle time Tc as follows:

$$\mathrm{Td} = \mathrm{Tc} - (\, 70/f_{\mathrm{clk}} + 1.8\ \mu\mathrm{s}\,)$$

where f_{clk} is the ADC0802 clock frequency. Td is the duration of the 555 output pulse. Td = 1.1RdCd; for a given Cd, Rd = Td/1.1Cd. A potentiometer can be used for Rd to allow

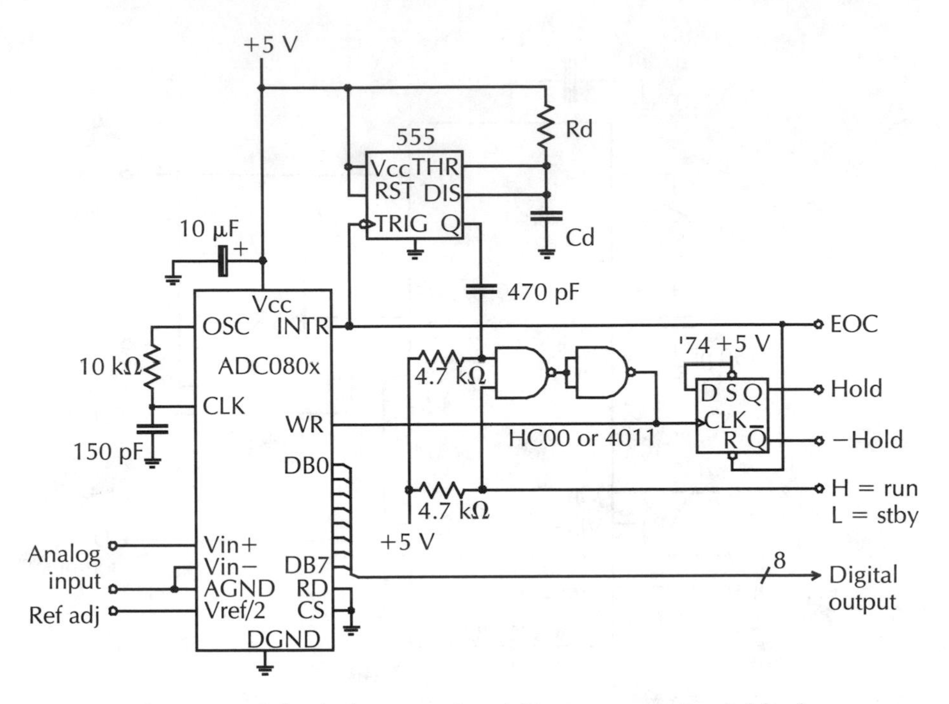

Fig. 6-26 A basic free-running A/D converter circuit block.

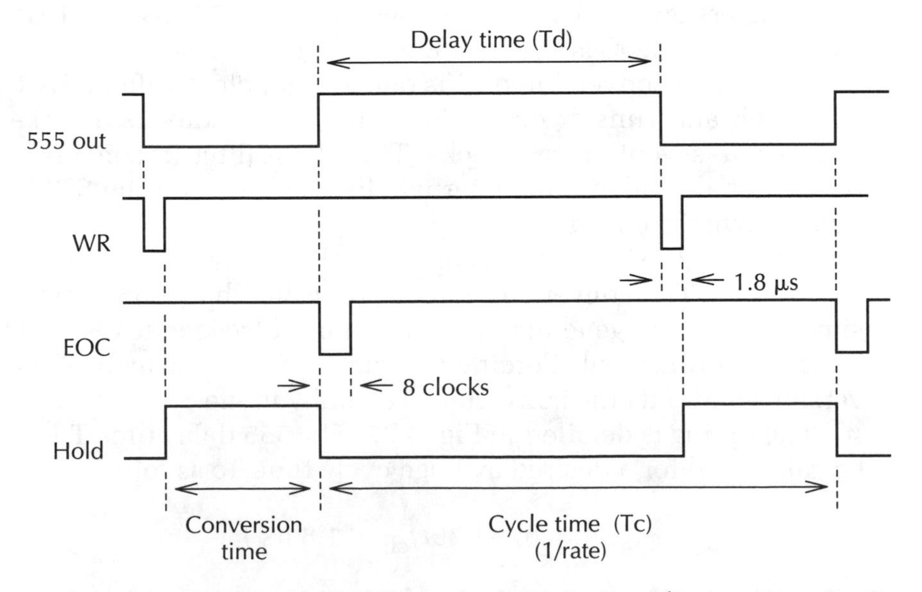

Fig. 6-27 Timing relationships in the free-running A/D converter circuit.

trimming the conversion rate. The average conversion rate with this simple circuit is reasonably stable, but if an exact rate is important, the external start version (Fig. 6-25) should be used.

Example: Suppose 2500 conversions per second are desired and the A/D clock is 600 kHz. Tc = $1/2500$ = 400 μs. Td = 400 – (70/600,000 + 1.8) = 282 μs. Let Cd be 10 nF. Then Rd = $(282 \times 10^{-6})/1.1(10^{-8})$ = 25.6 kΩ. Rd could be 22 kΩ in series with a 10-kΩ trimpot. (The average A/D conversion time is 70/600,000 = 117 μs.)

Scale factor and reference adjustment

As with all servo converters (see Fig. 6-22), the ADC0802 incorporates a D/A converter. The scale factor is set by the dc reference voltage applied to the internal D/A converter and is twice the reference voltage. The reference voltage is nominally 2.5 V, providing a converter input voltage range of 0 to 5 V.

A simple built-in reference voltage can be obtained from a resistor from the D/A to the +5-V Vcc supply. With this connection the scale factor will depend on Vcc. The D/A reference is also brought out to the $V_{ref}/2$ pin, which allows a more precise reference to be forced. A fairly stiff source such as an op amp should be used, as in Fig. 6-28. The input resistance at the $V_{ref}/2$ pin is about 1.2 kΩ.

Analog input circuitry

For best results, the input, +Vin, should always be driven directly by an operational amplifier. This can be a unity-gain buffer or the output of an antialiasing filter or track-and-hold.

The ADC0802 has quasi-differential inputs. The +Vin and –Vin are strobed sequentially so there is a 4½ clock period skew between them. This means that there is adequate common-mode rejection only at very low frequencies. For this reason it is best not to use the –Vin input at all. If a differential input is needed, a conventional operational amplifier differential input circuit should be used.

For accurate conversion of signals above a few hertz, a track-and-hold circuit should be used. There are many ICs available, such as the LF398. It is not difficult to build a satisfactory track-and-hold circuit. Several configurations are shown in Fig. 6-29.

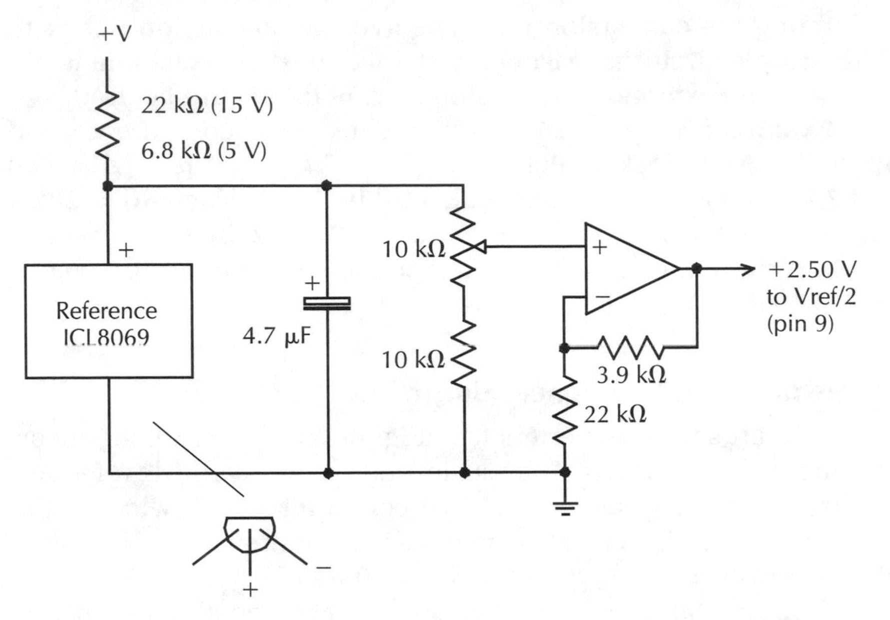

Fig. 6-28 *A circuit for trimming the ADC080x reference voltage.*

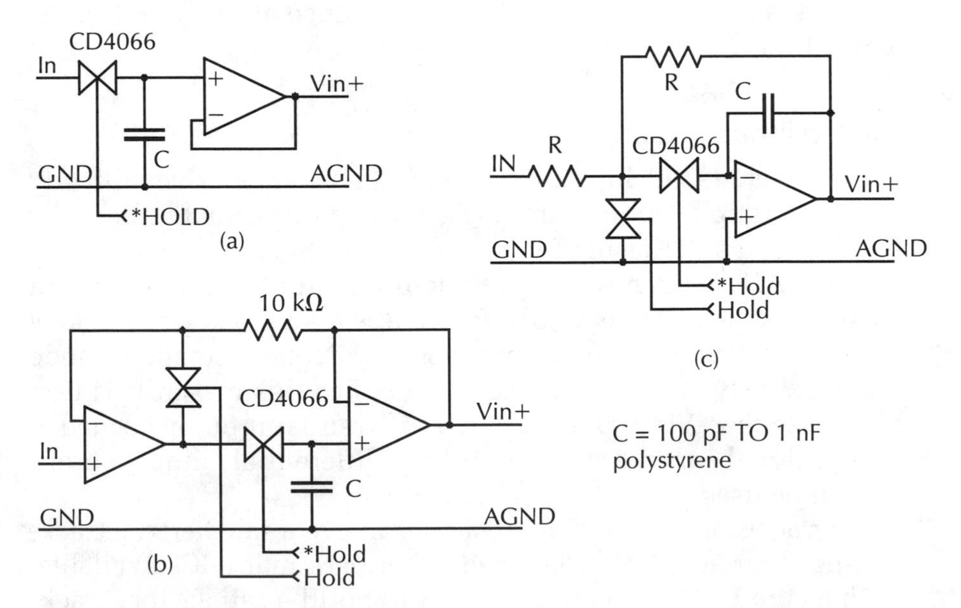

Fig. 6-29 *Track-and-hold circuits: (a) simple buffered hold; (b) closed-loop; (c) integrating.*

The simplest, circuit A, is just a storage capacitor and one section of a CD4066 CMOS switch with a buffer to isolate the capacitor. The circuit must be driven from a low-impedance source such as directly from an op amp. The acquisition time (neglecting the amplifier driving the circuit) depends on the channel resistance of the switch and the value of the storage capacitor; for 470 pF it would be 3 to 4 μs. An FET amplifier, such as the TL070 series, should be used for the buffer. A storage capacitor of 100 to 220 pF should be adequate for the applications discussed in this chapter. The capacitor must be a low-leakage type (not ceramic); polystyrene is recommended. Although this circuit might seem too simple to be good, note that there are commercial ICs that use it (for example, Analog Devices SMP-04).

Circuit B is the classic track-and-hold. By enclosing the switch in a feedback loop, the linearity during tracking and the acquisition time are improved. The second switch keeps the input amplifier in closed-loop condition during the hold period.

An integrator makes an excellent track-and-hold, as in circuit C. An advantage is that the switches are at virtual ground which minimizes the effect of their nonlinearity on the circuit. The switch connected to the junction of the resistors is not strictly necessary, but it does keep the input resistance of the circuit constant during the hold period. During tracking, the circuit forms a single-pole low-pass filter. For typical values of R = 10 kΩ and a 100-pF capacitor, the –3-dB frequency is 160 kHz. Note that the signal polarity is inverted.

Output interfacing

The A/D converter block can be combined with one of the serial or parallel I/O interface blocks described earlier in this chapter to make a compact and very useful data acquisition system.

Parallel output The parallel output block described earlier fits nicely with the converter block. (See A in Fig. 6-30). The data lines go to the output buffer. The strobe line connects to the converter block EOC output. The EOC pulse width is eight converter clocks, about 13 μs for a 600-kHz clock rate, which is satisfactory for use as is for the short parallel interrupt strobe (pulse stretch in Fig. 6-1 forced low).

Serial output A serial interface is also quite simple. The number of A/D conversions per second must be less than the se-

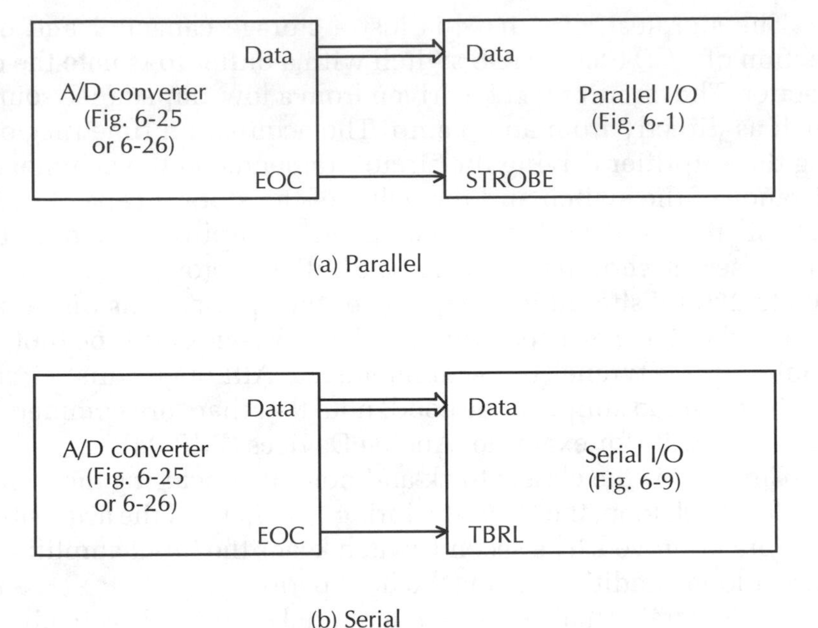

Fig. 6-30 *Output interface circuits for the A/D converter block: (a) parallel; (b) serial.*

rial byte rate (a tenth of the baud rate for eight-bit without parity or seven-bit with parity). If the serial transmission rate has been set to 38,400 baud, for example, the A/D converter must run at 3840 conversions per second or less.

An interface using the serial I/O block described earlier is shown in B of Fig. 6-30. As shown by the dotted line, you could steal a clock for the A/D converter from the baud rate generator.

For the sake of clarity, only transmission paths for data output are shown in the circuits in Fig. 6-30. Practical systems would likely use the receive path also for various control and status operations. An example of such is the digital voltmeter described later in this chapter.

D/A converter circuits

In comparison to A/D conversion, D/A conversion is a cinch. A simple way is to let each bit of a binary number control an electronic switch connected to a group of binary-weighted voltages or currents. Nearly all practical D/A converters (DACs) work this way.

There are several different ways to realize such a circuit. One good way is the R-2R resistive ladder network shown in

Fig. 6-31. The resistor network requires only two values of resistance of moderate value, which lends itself well to monolithic fabrication. The FET switches, one per bit, connect each tap either to ground or to the output. The network produces a binary-weighted array of currents so the output of the converter is an analog current representing the digital value.

This configuration is not just a good D/A converter: it is also a versatile circuit element. It is essentially a digitally controlled

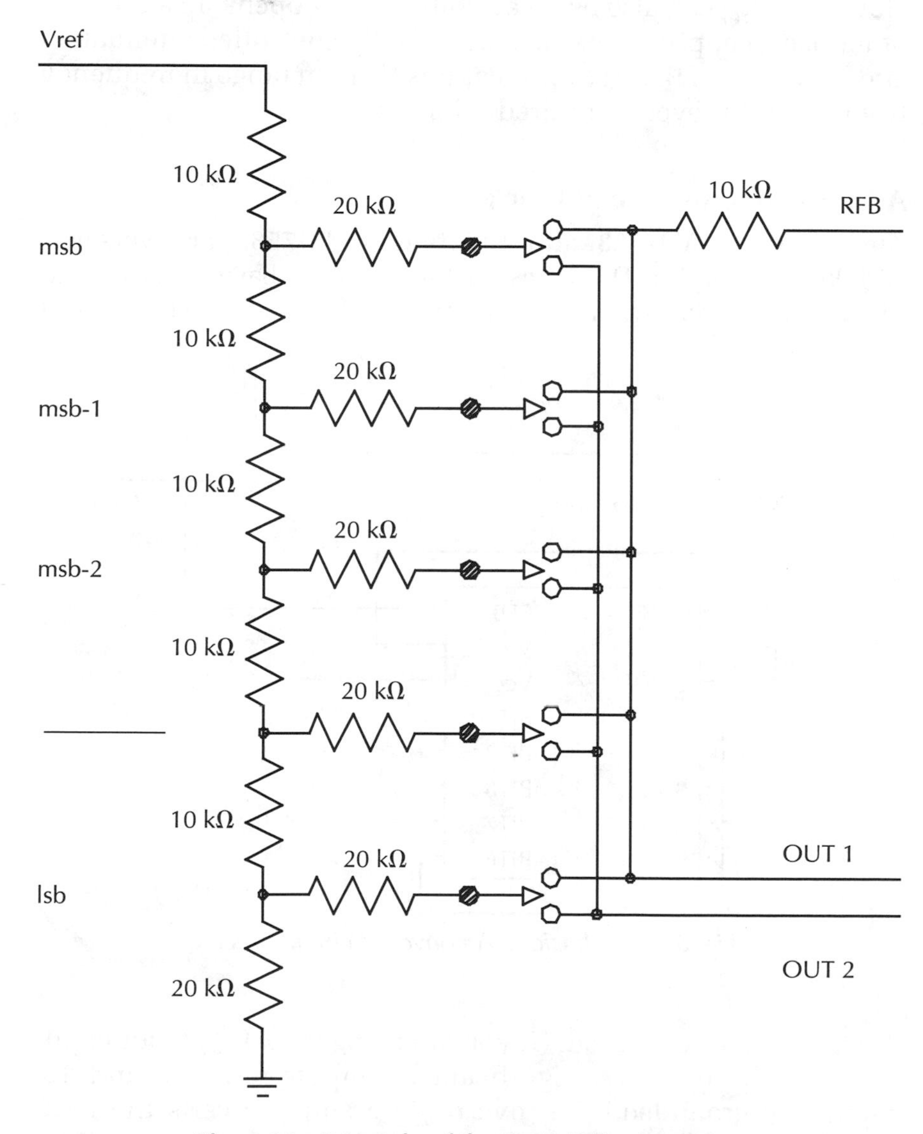

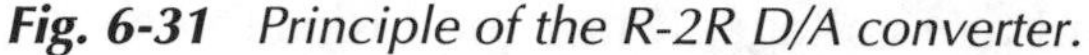
Fig. 6-31 *Principle of the R-2R D/A converter.*

potentiometer. The output current is a digital fraction of the current flowing through the resistor network— 0, $\frac{1}{256}$, $\frac{2}{256}$, . . . , $\frac{255}{256}$ of the current in the case of an eight-bit DAC. But the current through the network is proportional to the reference voltage V_{ref} applied to it. Both V_{ref} and the digital word determine the output current: in fact, the output current is their product. That's why this configuration is often called a *multiplying DAC.* When the device is used for D/A conversion V_{ref} will be a stable dc voltage. However, V_{ref} can also be an ac voltage. This opens up a range of other useful applications such as digitally controlled attenuators and multipliers. In these applications V_{ref} can range in frequency from dc up to several hundred kilohertz.

A D/A converter circuit block

The eight-bit AD7523 and the ten-bit AD7533 are versatile CMOS multiplying DACs, as in Fig. 6-31. A basic D/A circuit block using the AD7523 is shown in Fig. 6-32. The ten-bit

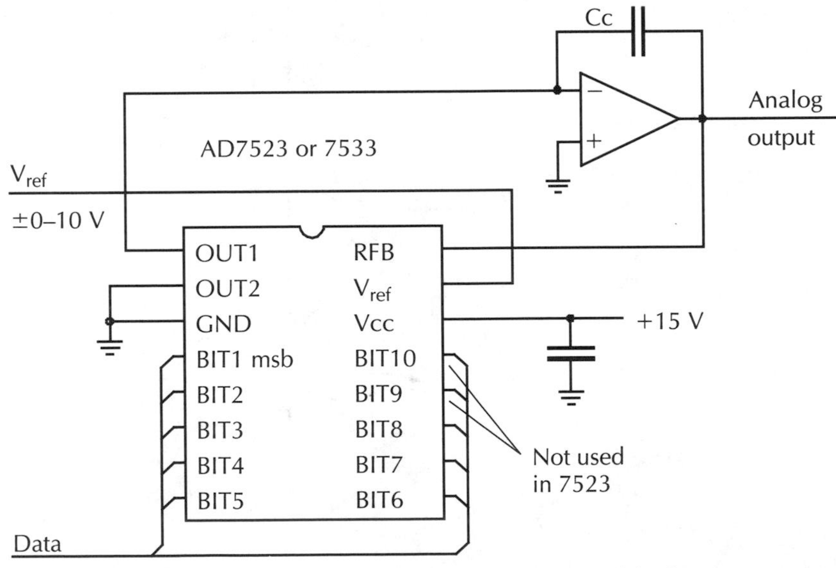

Fig. 6-32 *A basic D/A converter circuit block.*

AD7533 can also be used. (If you are using the AD7533 for eight-bit data, the two least significant bit inputs, pins 12 and 13, should be grounded.) An inverting op amp converts the DAC output current to a voltage. It is a good idea to use R_{FB} in the DAC as the amplifier feedback resistor whenever possible, because the actual resistance of a given part might differ by 30 percent or so from the nominal 10 kΩ. (The overall accuracy depends on the ratio of the ladder resistors.) Its temperature coefficient also matches the R-2R network. A small capacitor Cc compensates for the effect of the output capacitance of the DAC on the loop gain, providing better transient response. The value depends on the amplifier characteristics; 22 pF works well with most modern op amps such as the TL070 family. A larger capacitor can be used if low-pass filtering is desired.

Like most multiplying DACs, these devices are optimized for a ladder current of about 1 mA. The nominal input resistance of the ladder is 10 kΩ, so V_{ref} would be ±10 V in this case. Note that V_{ref} can be either positive or negative, or for that matter an ac voltage as when the DAC is used as a programmable attenuator. Because the op amp inverts, a negative V_{ref} gives a positive analog output voltage. When the DAC is used for straight D/A conversion, V_{ref} should be a stable voltage. Trimming the value of V_{ref} is the best way to set the overall scale factor precisely.

The built-in feedback resistor provides an overall scale factor of unity; that is, the analog output from the amplifier will go from zero to V_{ref} (actually 0 to $-(255/256)V_{ref}$). For other scale factors an external resistor Rf can be used. For a given full-scale output voltage Vo, Rf = (Vo*10 k)/V_{ref}.

Reconstruction filtering

The DAC output at each instant is the product of V_{ref} and the digital input. (There is actually a short time, less than 200 ns for these parts, for the output to settle to a new value after a change in the digital input value.) When the DAC is used for D/A conversion, the digital input will usually change with each new sample, remaining constant until the next sample. As pointed out previously, this produces a sin(x)/x filtering effect in the reconstruction. The best solution is oversampling, but for less demanding applications it might be sufficient to use a reconstruction filter with a response such that the overall response is nearly flat.

An example of such a filter is shown in Fig. 6-33. The solid curve is the overall frequency response; the dashed curve is the sin(x)/x response for comparison. Because the sin(x)/x response is taken into account in the filter alignment, the cutoff frequency f_C of this circuit cannot be freely selected but must be at the Nyquist frequency F_N; that is, at $F_S/2$, half the sampling frequency. The filter consists of two sections: a single real pole at

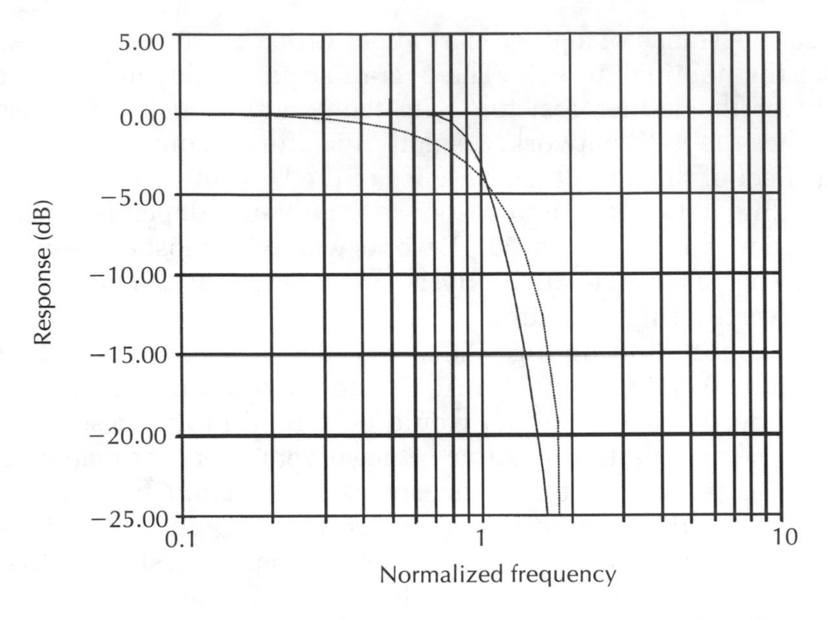

Fig. 6-33 *Effective frequency response of reconstruction filter with sin(x)/x compensation.*

$1.28F_N$ and a pole pair at F_N with damping α of 0.75.

One possible realization is the circuit in Fig. 6-34. For simplicity a single-amplifier configuration is used for the two-pole section. This section has a low-frequency voltage gain Ho of 2.25. An Rf of 4.7 kΩ for the current-to-voltage converter gives an overall scale factor of 10.57-V output for 1-mA full-scale DAC output; V_{ref} can be trimmed to set the output voltage exactly. The current-to-voltage converter is a convenient place to realize the single pole. Because Rf is already determined, the value of C1 is constrained to be $1/(2\pi)(1.28\ F_N)(4700)$. The value of C2 in the two-pole stage can be selected freely, and $R2 = 1/2\pi F_N C2$.

The general design equations for a desired analog output voltage Vo are as follows. R_{DAC} is the actual DAC ladder resistance of the part used, which is nominally 10 kΩ, but might vary ±30 percent or so from one part to the next. Then,

$$Rf = (Vo*R_{DAC})/(2.25\ V_{ref})$$
$$C1 = 1/2\pi(1.28\ F_N)Rf$$
$$C2 = 1/2\pi F_N R2$$

A digital voltmeter

This chapter concludes with a digital voltmeter (DVM) module—a subsystem, really—that ties together the various parts of the

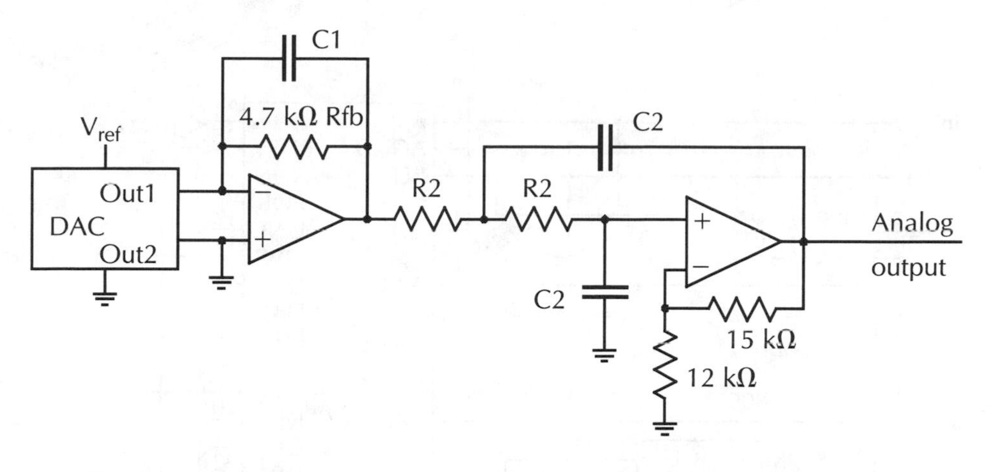

Fig. 6-34 *A third-order reconstruction filter with sin(x)/x compensation.*

discussion so far, and is also a working and useful unit in its own right. As a matter of fact, given a modicum of care in construction it is the equal of far more expensive commercial equipment. It can be used as a stand-alone voltmeter, as a high-resolution A/D converter module with BCD output, or as a versatile front end to a PC data acquisition system.

The basic DVM provides ±2-V full-scale sensitivity with 4½ digit resolution. The least significant digit (lsd) thus indicates increments of 100 μV. As with all digital measurements of this sort, there is a ± 1 lsd uncertainty in the reading.

A word about the uncertainty (quantizing noise) in all digital measuring devices seems in order. The uncertainty is largely random, so the true value can be approximated by averaging several successive measurements. This will also reduce any random noise in the signal being measured. The improvement in resolution and signal-to-noise ratio goes as the square root of the number of measurements. Averaging ten sequential values, for example, improves the ratio by 3.16 (10 dB). Unfortunately certain kinds of low-frequency noise—so-called 1/f noise—is chaotic, not random, and such noise does not converge to a true value under averaging.

The 4½-digit A/D converter

The heart of the DVM, Fig. 6-35, is the Harris-Intersil ICL71C03 and ICL8068 chip set. Together they provide a complete 4½-digit BCD A/D converter. The 8068 contains the analog circuitry (integrator and comparator) and the 71C03 contains the digital circuits and counter.

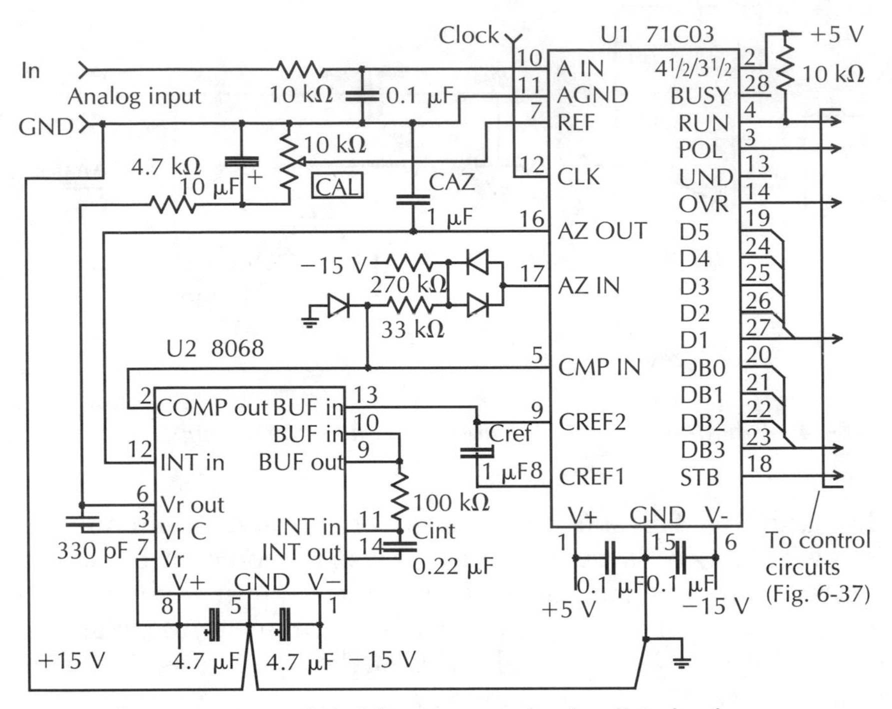

Fig. 6-35 *A 4½-digit A/D converter for the digital voltmeter.*

The circuit block in Fig. 6-35 is intended to be combined with other circuit blocks (to be described shortly) in order to form a complete voltmeter. You can readily customize it to your particular needs. You can choose from among the several supporting modules described here or work up your own. The minimum would be a clock generator and an analog input circuit. Note that there is no dc return resistor on the analog input; if an input circuit such as in Fig. 6-42 (shown later) is not used, an input resistor (typically 1 MΩ or 10 MΩ) should be added between analog input and analog ground.

The chip set forms a dual-slope A/D converter that measures the ratio of an input voltage to a reference voltage. The principles were outlined earlier. The conversion cycle is like the one depicted in Fig. 6-21 except that an autozero phase has been added. During autozero the integrator input is grounded and its output voltage is stored in capacitor C_{AZ}. This voltage is then used as the integrator zero reference during the measurement phases, neatly eliminating amplifier offsets and other zero errors. The autozero phase takes 10,001 clocks. It is followed by the signal integration phase, 10,000 clocks. The reference integration phase is exactly twice as long, 20,000 clocks (plus 1 clock not included in the count), giving a scale factor of 2 so that a reference voltage of 1.0000 V gives an input range of ±2.0000 V. The total conversion time is 40,002 clocks.

Although the accuracy of a dual-slope A/D converter does not depend on the actual values of the clock or integrator (they need only be constant during a conversion cycle), there are constraints in practical circuits in order to maximize dynamic range and linearity. The component values shown are for ±1.9999-V full scale, and for a 120-kHz clock that gives three conversions per second.

Data output from the 71C03 is in a multiplexed BCD format. A data strobe output, *STB, strobes each digit in sequence, D5 (msd) to D1 (lsd), at the end of each conversion. The strobe width is ½ clock (4.2 μs). There are five digit drive signals D5–D1 for controlling a display. The drive pulses are 200 clocks wide (1.66 ms) and each digit is refreshed every 1000 clocks (8.33 ms or 120 Hz), a duty cycle of 20 percent.

The 71C03 has three flag outputs: POLarity is high for positive inputs and low for negative; OVR is high for overrange values (outside of ±19,999); and the UNDERflow flag goes high for values less than 9 percent of full-scale (that is, under ±1800) and returns to low when the next measurement cycle begins. OVR and UNDER can be used to control autoranging circuits. There is also a status output, BUSY, which is high during a conversion (the signal and reference integrate phases).

There are two control inputs. With RUN/HOLD at high level the converter runs continuously. If RUN/HOLD is brought low, the converter enters a standby mode at the end of the current conversion. Data strobes are not emitted during standby, but the digit drives and data output remain active so a readout will continue to display the last measurement. When RUN/HOLD returns to high level, conversion cycles resume. A low level on the 4½/3½ input short-cycles the converter, forcing a 3½-digit resolution.

Clocking Circuit A in Fig. 6-36 is a simple 120-kHz clock oscillator using CMOS inverters. If the basic DVM is to be used with a serial interface and the baud rate generator, you can simply help yourself to the 1.2288-MHz master oscillator and divide it by 10, as in circuit B.

Reference voltage The ICL8068 has a built-in reference voltage regulator with a stability on a par with common digital multimeters. For greater accuracy and stability a precision reference

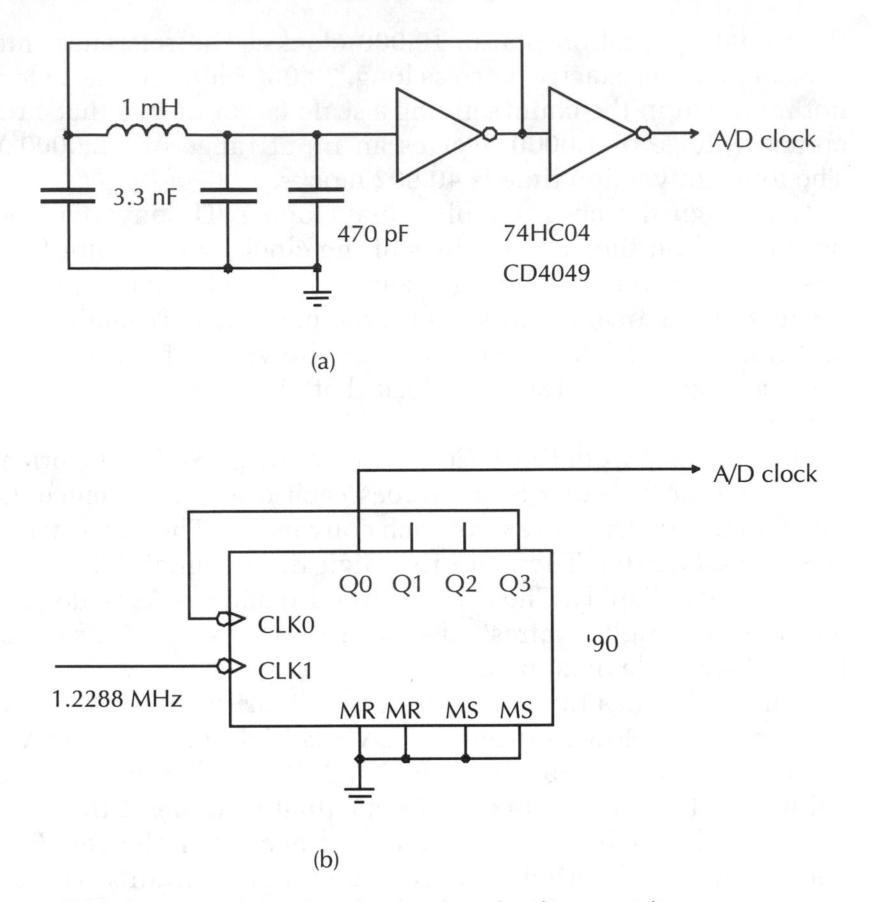

Fig. 6-36 *Clock circuits for the digital voltmeter A/D converter.*

can be substituted. The full-scale reading will be twice the reference voltage; for ±2-V full scale, the reference is +1.0000 V.

The DVM will have to be calibrated before use. The best way is with a standard cell or a precision calibration source. Another good way is to calibrate it against another meter of known accuracy using a stable dc voltage such as an ordinary 1.5-V dry cell. If neither a calibration source nor another meter is available, a dry cell can be used. The nominal voltage of a fresh and unused carbon-zinc dry cell is 1.55 to 1.56 V.

Control and data formatting

By adding a few supporting circuits the flexibility and usefulness of the DVM core circuit is much enhanced. Figure 6-37 shows the support and control circuits.

The raw output is five four-bit BCD values. They could be

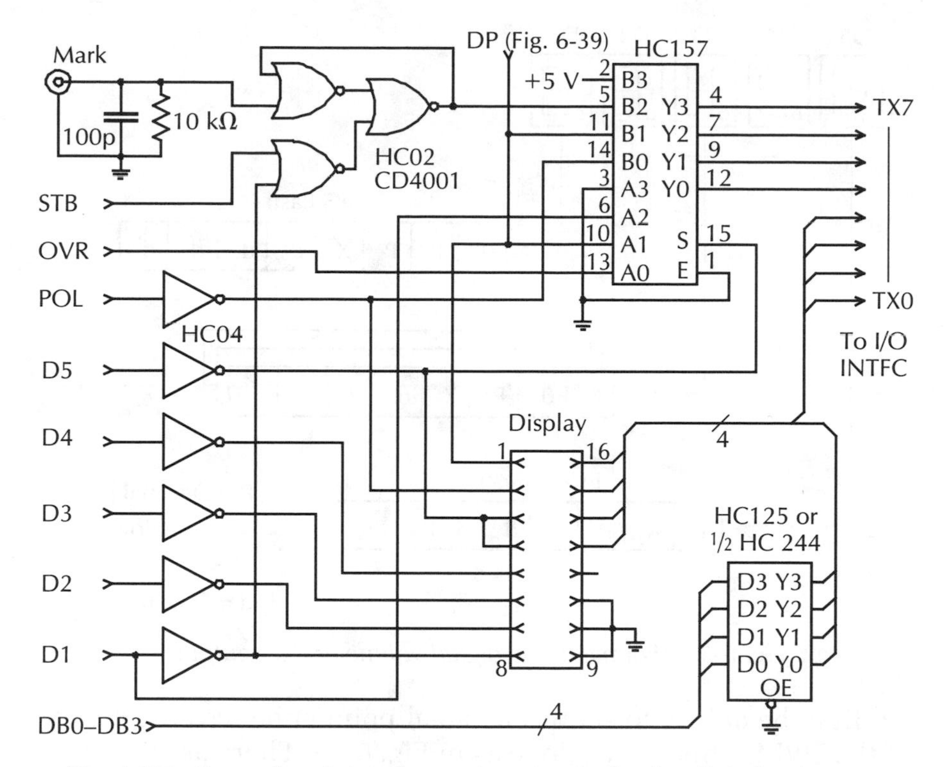

Fig. 6-37 *Control and data formatting circuits for the digital voltmeter.*

sent to a computer or other destination as is, or packed two to a byte, but a better idea is to piggyback some additional information on the BCD data using the upper four bits of each byte. Polarity, an overflow flag, a decimal point flag, a marker, and identification of first and last byte are added. The bit format of each of the five bytes sent to the PC after each conversion is shown in Fig. 6-38. Note that the A/D polarity flag has been inverted and renamed MINUS; the minus bit in the data output is zero for positive values and 1 for negative.

In many applications it's desirable or necessary to be able to add an event marker to the data, so a simple event marker latch is included. (Latching allows short pulses to be used to set a marker.) The event marker input accepts a positive CMOS-level pulse. At the end of the current conversion cycle the marker bit in the output data is set and the last digit strobe resets the marker latch.

The digital output of the A/D converter is an integer value, −19,999 through +19,999, expressed in BCD-plus-sign format.

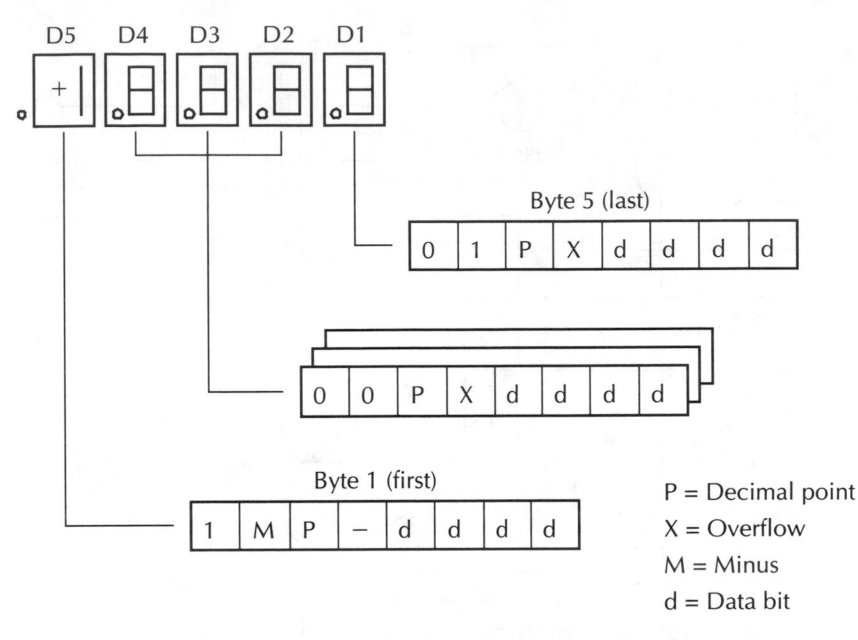

Fig. 6-38 *Format of the digital voltmeter data output.*

Adding the ability to show a decimal point enhances the utility of the DVM. This can be done as in Fig. 6-39. There are five decimal point control lines, one for each digit. Pulling one of the lines low sets a flag bit in the data byte and also lights the decimal point on the display. If input processing circuits are added to or used with the basic DVM, extra contacts on a range switch could control the decimal point location.

Table 6-4 Decimal point control bits.

Bit 4	3	2	Digit	Readout
0	0	0	None	N N N N N
0	0	1	D1 (lsd)	N N N N .N
0	1	0	D2	N N N .N N
0	1	1	D3	N N .N N N
1	0	0	D4	N .N N N N
1	0	1	D5 (msd)	.N N N N N
1	1	0	None	N N N N N
1	1	1	None	N N N N N

There is also provision for remote control of the decimal point and the RUN/HOLD function of the DVM from the computer. Bit 0 controls run or hold mode. Setting it to 1 puts the

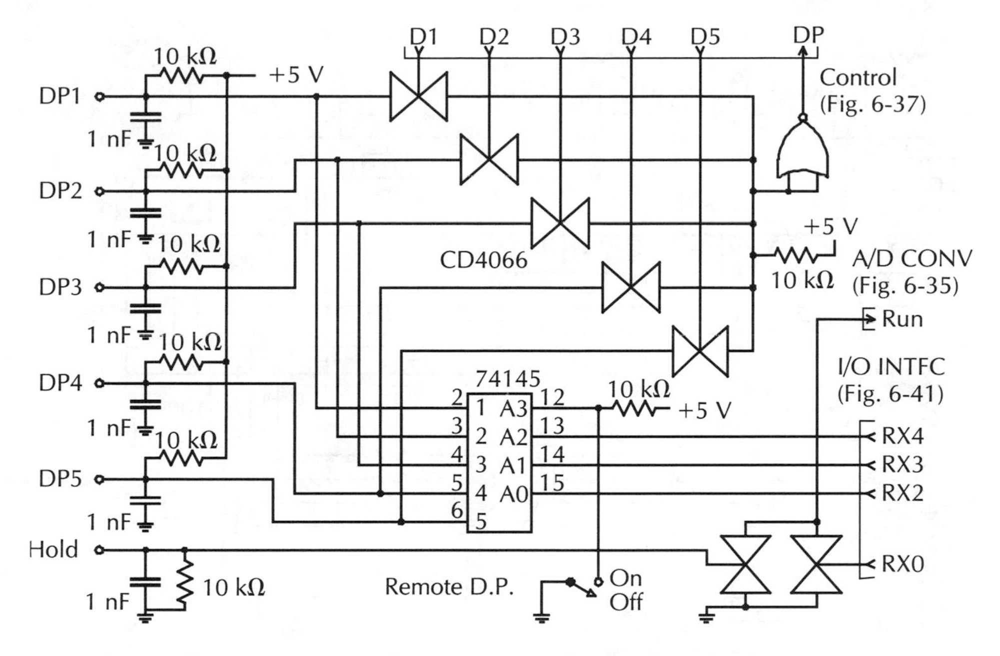

Fig. 6-39 *Decimal point circuits for the digital voltmeter.*

converter in hold mode. Bits 2,3, and 4 are a three-bit binary nibble that selects the decimal point as shown in Table 6-4. A control byte value of 0 selects default operation, namely run mode with no decimal point. The DVM interface software (Fig. 6-45 shown later) shows a way to implement a control byte. The remaining bits of the control byte are not assigned and you could use them to implement other control functions.

Remote control of the decimal point lines is effected by a '145 open-collector BCD to 1-of-10 decimal decoder. Decimal outputs 1 through 5 are tied to decimal point lines DP1 to DP5 respectively. A slight trick is used to enable and disable remote decimal point control. Setting the most significant bit input of the '145 high forces decodes to be greater than or equal to 8 and hence never 1 through 5.

Display circuits

The control section in Fig. 6-37 shows a connector for a readout. A simple LED display module is shown in Fig. 6-40. The display is optional and because it draws up to 270 mA you might want to omit it in the case of battery operation to reduce power consumption. (Liquid-crystal [LCD] displays use very

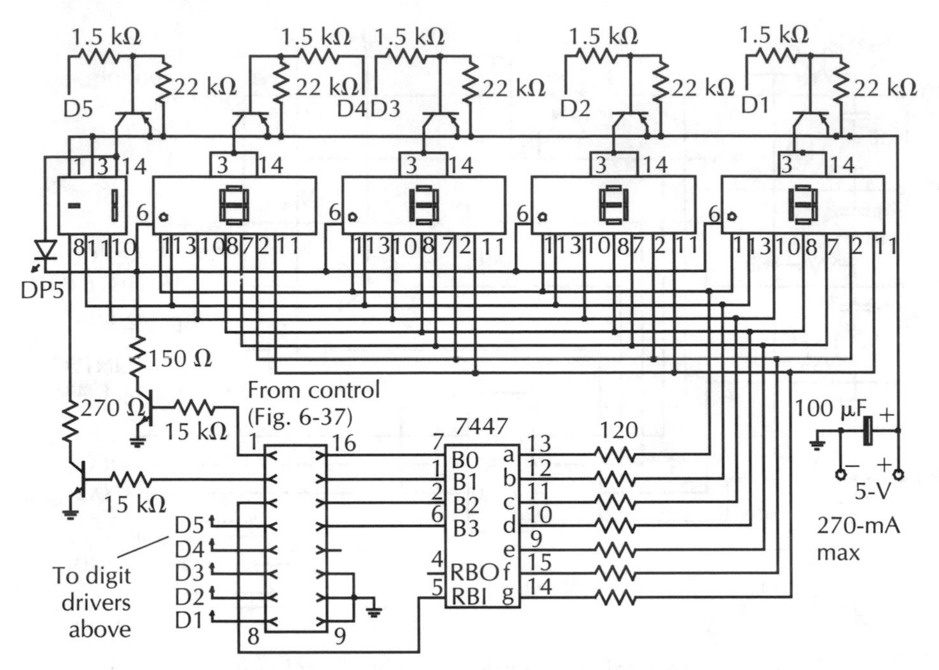

Fig. 6-40 LED display for the digital voltmeter.

little power and could be used to provide a display with battery operation.) A sixteen-pin DIP connector is convenient for the readout data connector.

The readout module in Fig. 6-40 is conventional and uses common-anode LED displays. D4 through D1 use seven-segment displays with a left-hand decimal point (for example, the MAN72A whose pinout is shown in the figure). D5 is a ±1 (or "overflow") display such as the MAN73A. Because most ±1 displays don't have a decimal point a separate small LED is used, placed to the left of the D5 display IC. Because of the high supply current a separate connection is used for the 5-V supply. It should go directly to the power supply output terminals, not to any of the voltmeter circuit boards.

I/O interface circuits

Either the serial (Fig. 6-9) or the parallel (Figs. 6-1 and 6-2a) interface blocks can be used with the DVM for data transmission to a PC. The connections are straightforward, as illustrated in Fig. 6-41. The serial interface shows the voltmeter A/D converter clock being derived from the baud generator oscillator.

It is possible to provide both serial and parallel interfaces in

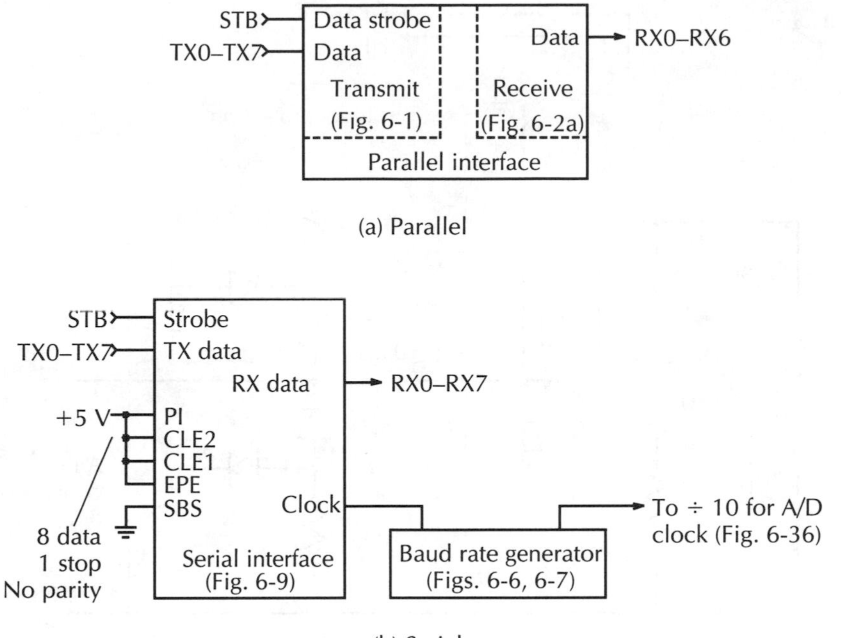

Fig. 6-41 *I/O interfaces for the digital voltmeter.*

one unit. The transmission sections of both interfaces can be active simultaneously. Only one interface at a time can be used for remote-control functions, however, so the receiver sections need to be modified slightly. The three-state enable lines of the receive buffers in both units must be controlled. In the serial interface (Fig. 6-9) this is RRD (pin 4) of the 6402 UART, and in the parallel interface (Fig. 6-2a) it is pins 1 and 19 of the '244 buffer. The buffer output enable line of either the serial or parallel interface is brought to low level to enable it. The enable line of the other interface must remain high (not enabled).

Input circuits

You now have a complete DVM subsystem. Although it could be used alone, its fixed ±2-Vdc input voltage range is intended for use in a measuring setup that would provide a "front end" with range selection and other input signal processing. There are many kinds of input circuits that could be used to provide many kinds of measurement functions. To suggest some of the possibilities, here are two examples.

Figure 6-42 shows a dc range selector giving ±2-V and ±20-V ranges, well suited to the 1-V and 10-V full-scale conven-

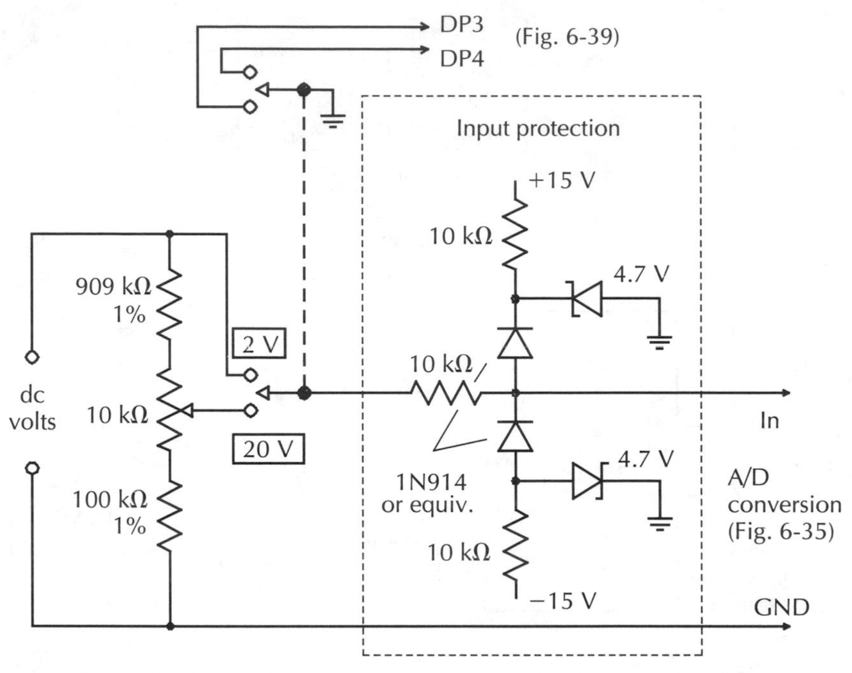

Fig. 6-42 *Example of a dc input circuit for the digital voltmeter.*

tions in analog sensor technology. Note that a second set of contacts on the range switch is used to set the decimal point. The input resistance is 1 MΩ, the customary value for oscilloscopes and similar instruments. (Multiplying all of the resistance values by ten will provide the 10-MΩ input resistance commonly found in voltmeters.) A small potentiometer is included to trim the divider because the resolution of the DVM is much greater than 1 percent resistor tolerance. Calibration of the divider is simple. A stable dc voltage, such as a 1.5-V dry cell, is measured on the 2-V range. Then on the 20-V range, the potentiometer is adjusted for a reading of exactly one-tenth of that value.

Figure 6-42 also shows a method for overvoltage protection. A pair of reverse-biased small-signal silicon diodes (such as the 1N914) clamp the input to a maximum of about ±5.3 V. Excess input current during overload is shunted away through the zener diodes. Including such a circuit in your implementation is strongly recommended. Although the 71C03 can tolerate occasional slight overloads, substantial overvoltage can damage the chip.

The ac voltages can be measured with the circuits in Figs. 6-43 and 6-44. Together they provide a first-rate ac front end. The wideband input noise is less than 10 μV rms and the frequency

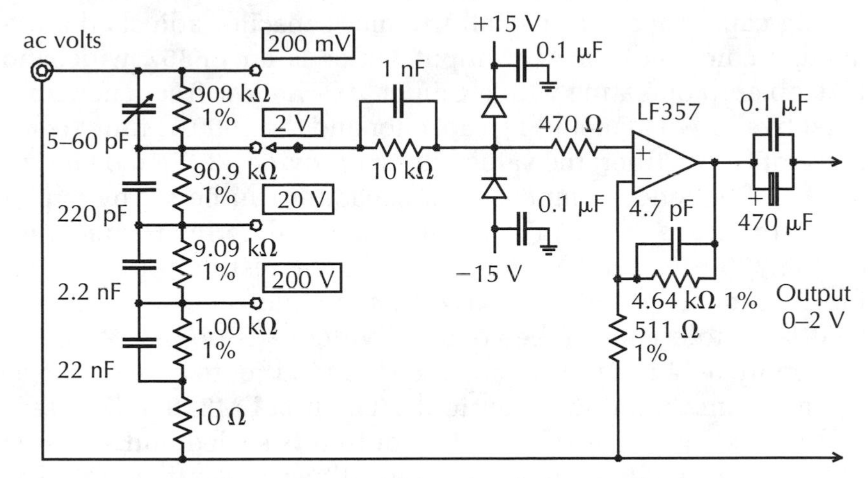

Fig. 6-43 *A wideband ac input circuit.*

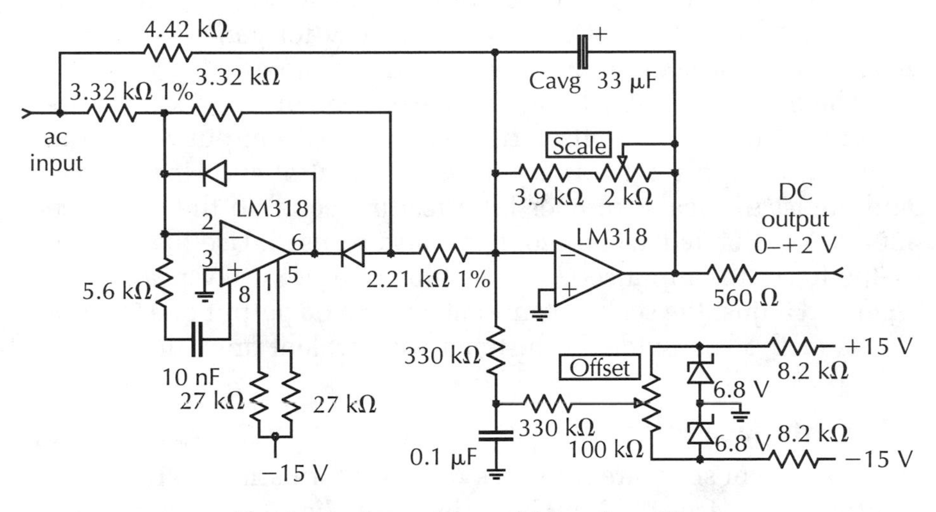

Fig. 6-44 *A precision wideband rectifier.*

response is flat within ±0.05 dB from 5 Hz to 250 kHz. This is substantially better performance than the ac voltage function in the average commercial multimeter. As a matter of fact, you could use these two circuit blocks as a front end to improve the performance of other voltmeters you have.

Figure 6-43 is a range selector and 20-dB input amplifier. The range switch is frequency-compensated and provides several voltage ranges from 200 mV to 200 V. The input resistance is 1 MΩ and the input capacitance is approximately 30 pF plus any stray wiring capacitance. The small trimmer capacitor adjusts the frequency compensation. The input buffer is an LF357 wideband FET op amp operating with a gain of 10. The low-frequency limit is set by the 470-μF coupling capacitor and the 1900-Ω input resistance of the rectifier; the values shown provide –3 dB at 0.18 Hz.

The frequency-compensation capacitor can be set by applying a sine wave of 100 mV or so at 1 kHz and noting the readings on the 200-mV and 2-V ranges and then adjusting the capacitor for the same readings at 50 kHz (this method cancels out the effect of the tolerance of the voltage divider resistors).

Figure 6-44 is a precision wideband rectifier to convert the ac signals from the input buffer to dc. The first LM318 amplifier is really a half-wave rectifier, but its output is scaled and summed with the unrectified input to give full-wave rectification. The second op amp is a summer and averaging (filtering) circuit. As with all such circuits, the dc output is proportional to the average absolute value, not the "true rms." The averaging time constant is about 160 ms which is satisfactory for general-purpose use; it can be increased by increasing the value of C_{avg}.

The circuit is best calibrated against an ac meter of known accuracy. First, adjust the offset trimmer for zero dc output with no ac input. Then apply a 1-V rms sinusoidal input of 400 Hz to 1 kHz and adjust the scale trimmer for a reading equal to the reference meter. A calibrated oscilloscope can also be used. Use the oscilloscope to set the input source to a value of exactly 1-V peak (or 2-V_{PP}) and adjust the scale potentiometer for a dc output of 0.7071 V. These methods provide the customary equivalent rms calibration.

DVM interface software

The low-level software interface routines for using the DVM are contained in a unit, DvmUnit, listed in Fig. 6-45. These routines assume that the DVM is using a serial interface, and draw on the Serial Toolkit from chapter 4. If you are using the DVM with a parallel interface, the routines are easily changed to use the Parallel Toolkit from chapter 5. The functions of most of the routines in the serial and parallel toolkits are (purposely) the same.

The DVM sends out data in five-byte sequences. The routine GETVAL returns each such sequence in a little eight-byte buffer

Fig. 6-45 *The DVM unit: utility routines for the digital voltmeter.*

```
{ ------------------------------------------------------------------------- }
{                               DVM UNIT                                    }
{     A Library of Input and Control Routines for the Digital Voltmeter     }
{ ------------------------------------------------------------------------- }
(* DVMUNIT.PAS 1.1 ©1992 J H Johnson *)


UNIT
   DvmUnit;

INTERFACE

USES
   SerialTk;

TYPE
   DvmBlock = array [0..7] of byte;
   DvmStr = string[8];
   DvmRec = record                       { holds the info for each DVM reading }
               Mag : integer;
               DP : byte;
               Mark,
               Ovr : boolean;
            end;

CONST
   dvmNoInput = 1000;                                            { error code }
   TimeoutCount = 182;               { timeout at 10 sec (18.2 ticks * 10 sec) }

VAR
   TestType : (tmSine,tmRand,tmOvr);                       { kind of test signal }


procedure GetVal( var V : DvmBlock; var E : integer );
    { gets bytes from DVM, returns as array D5..D1 with flags and BCD values }

procedure ParseDvm( V : DvmBlock; var R : DvmRec );
    { fill DvmRec with value and flags }

function DvmValStr( V : DvmBlock ) : DvmStr;
    { returns ASCII string with value only }

procedure SetDvmHold( Hold : boolean );
    { TRUE to hold, FALSE to run }

procedure SetDvmDP( DP : byte );
    { set left-hand decimal point on digit DP (5 is msd) }

procedure TestGetVal( var V : DvmBlock ; var E : integer );
   { synthesize an array of bytes in DVM format for testing }

IMPLEMENTATION
{ ========================================================================= }

VAR
   Ticks : longint absolute $40:$006C;     { system "tick" value at 40h:006Ch }
   OpFlags : byte;
```

Fig. 6-45 Continued.

```
{ ---------- input and data formatting ---------- }

{ gets bytes from DVM, returns as array D5..D1 with flags and BCD values }

procedure GetVal( var V : DvmBlock; var E : integer );
var
   B,F,P : byte;
   T : longint;
begin
   E := 0;
   P := 0;
   T := Ticks + TimeoutCount;
   repeat
      if RxDataWaiting then
         begin
            B := ReadSerial;
            F := B and $C0;                    { top 2 bits into flag variable }
            if F >= $80 then P := 0;                             { first byte }
            V[P] := B;
            P := (P + 1) and 7;
         end;
      if Ticks > T then
         begin
            E := dvmNoInput;
            Exit;
         end;
   until F = $40;                                              { got last byte }
end; {procedure GetVal}


{ extract and format info from V, a raw DVM byte string }

procedure ParseDvm( V : DvmBlock; var R : DvmRec );
begin
   R.DP := 0;
   R.Mark := ( V[0] and $40 <> 0 );
   if V[0] and $20 <> 0 then R.DP := 5;
   if V[0] and 1 = 1 then R.Mag := 10000 else R.Mag := 0;
   R.Ovr := (V[1] and $10 <> 0);
   if V[1] and $20 <> 0 then R.DP := 4;
   R.Mag := R.Mag + ( (V[1] and $0F) * 1000 );
   if V[2] and $20 <> 0 then R.DP := 3;
   R.Mag := R.Mag + ( (V[2] and $0F) * 100 );
   if V[3] and $20 <> 0 then R.DP := 2;
   R.Mag := R.Mag + ( (V[3] and $0F) * 10 );
   if V[4] and $20 <> 0 then R.DP := 1;
   R.Mag := R.Mag + (V[4] and $0F);
   if R.Ovr then R.Mag := 20000;
   if (V[0] and $10 <> 0) then R.Mag := R.Mag * -1;
end;


{ make a character string from a raw DVM byte string }

function DvmValStr( V : DvmBlock ) : DvmStr;
var
   I : byte;
   S : DvmStr;
begin
```

Fig. 6-45 Continued.

```
   if (V[0] and $10 <> 0) then S := '-' else S := '';
   I := 0;
   repeat
      if (V[I] and $20 <> 0) then S := S + '.';                { decimal point }
      S := S + chr( (V[I] and $0F)+48 );
      if (V[I] and $C0 = $40) then I := $FF else inc(I);
   until (length(S) = 8) or (I = $FF);
   DvmValStr := S;
end;


{ ---------- DVM control ---------- }

procedure SetDvmHold( Hold : boolean );
begin
   if Hold then OpFlags := OpFlags or 1 else OpFlags := OpFlags and $FE;
   WriteSerial( OpFlags );
end;

procedure SetDvmDP( DP : byte );
begin
   OpFlags := (OpFlags and $E3) or ((DP and 7) shl 2);
   WriteSerial( OpFlags );
end;


{ ---------- routine to synthesize dummy input for testing ---------- }

procedure TestGetVal( var V : DvmBlock ; var E : integer );
const
  TwoPi = 6.2831853;
  Omega : real = 0.0;                              { sine argument; static var }
var
   N : real;
   Nr,Mag : integer;
   K,Over : byte;
begin
   if TestType = tmRand then
      begin
         { random }
         Nr := Random( 20000 );
         Nr := Nr - 10000;
      end else
      begin
         Omega := Omega + 0.0057;
         if Omega > 1.0 then Omega := 0;
         N := sin( TwoPi * Omega );
         Nr := trunc( 10000 * N );
      end;
   if TestType = tmOvr then Nr := Nr * 3;
{ make fake DVM string }
   Mag := abs( Nr );                                   { use absolute value }
   if Mag > 19999 then Over := $10 else Over := 0;          { overflow flag }
   V[0] := (Mag div 10000) or $80;               { upper nibble = first digit }
   if Nr < 0 then V[0] := V[0] or $10;       { also set minus bit if negative }
      Mag := Mag mod 10000;                               { leave remainder }
   V[1] := (Mag div 1000) or Over;
      Mag := Mag mod 1000;
```

Fig. 6-45 Continued.

```
   V[2] := (Mag div 100) or Over or $20;              { set dec pt on 3rd digit }
      Mag := Mag mod 100;
   V[3] := Mag div 10 or Over;
   V[4] := (Mag mod 10) or Over or $40;             { upper nibble = last digit }
{ test marker if using random }
   if (TestType = tmRand) and (Nr > 9900) then V[0] := V[0] or $40;
   E := 0;
end; {of TestGetVal}


BEGIN
   OpFlags := 0;                               { preset to Run, no decimal point }
   TestType := tmSine;
END.
```

of type *DvmBlock*. The measurement data from the DVM is in BCD code with added status bits, so some manipulation is needed to put the information into a more usable form. The routine **ParseDvm** does this, putting the results into a record of type *DvmRec*. The routine **DvmValStr** translates a measurement into a literal string for use where a string of numerals is needed, for display on the screen for instance.

Extracting a numerical value from the DVM output is thus a two-step operation. The manipulations in **GetVal** and **ParseDvm** could have been combined in a single faster routine, but speed is hardly a problem at the rate of three measurements per second and keeping the routines separate makes them a little more flexible. For example, the **DvmValStr** routine can take advantage of the close relation of BCD and ASCII to build a string of characters directly from the raw DVM data very efficiently. It is also true that by using assembly language and taking advantage of the fact that the 80*x*86 performs the mod and div operations simultaneously, the parsing of the raw data could have been implemented more efficiently. Here, too, speed is not very important and it was felt that the clarity of writing in a high-level language outweighed considerations of efficiency. This underlines the fact that in programming the "best" approach depends on the circumstances.

The subroutines in the DVM unit are as follows:

- **GetVal** returns the five-byte sequence the DVM sends for each measurement. The routine repeatedly fetches bytes until the first byte of a measurement (digit D5) is encountered and then fetches the remaining four bytes (digits D4 to D1). This ensures proper framing of the sequence. If a valid measurement is not found within a certain time, the routine quits and returns the error code *dvmNoInput*. The system "ticks" are monitored and the value assigned to *TimeoutCount* determines the maximum time. The initialization code of the unit presets the value to 10 seconds (182 ticks).

- **ParseDvm** parses the raw BCD measurement string, converts the BCD values to binary, extracts the flags and decimal point, and fills a record of type *DvmRec* with the results. The decimal point value in *DvmRec.DP* is 1 to 5 for digit D1 to D5, or 0 for none.
- **DvmValStr** takes a raw BCD measurement string and returns the measured value as an ASCII string.
- Calling **SetDvmHold** with the value true sets the DVM to hold mode. A value of false sets run mode. (The DVM control switches must be set so that remote control of run/hold is enabled.)
- **SetDvmDP** sets the decimal point indicator in the DVM. Calling it with a value of 1 sets the point on digit 1 (the lsd), 2 sets digit 2, and so on. A value of 0 turns off the decimal point indicator in the DVM. (The DVM control switches must be set so that remote control of decimal point is enabled.)
- **TestGetVal** is included for testing analysis and display software without having the DVM connected. It works just like **GetVal** except that it synthesizes the measurement. The kind of data generated depends on the current value assigned to the variable *TestType*. The value *tmSine* generates a sine wave; *tmRand* simulates random noise; and *tmOvr* simulates occasional overflow. The initialization code of the unit sets the value to *tmSine*. (The sample data acquisition program DvmScope in chapter 7 shows how this routine can be used.)

The small demonstration program DVMDEMO in Fig. 6-46 shows how these routines can be used with the DVM system. This little program is also useful for testing the DVM. It is assumed that the DVM uses a serial interface; if a parallel interface is used, the baud rate selection would be replaced by a parallel port selection.

DVMDEMO doesn't do anything with the data from the DVM; it just displays the results on the screen as a column of numbers.

Fig. 6-46 *Demonstration and test program for the digital voltmeter.*

```
{ ------------------------------------------------------------------------- }
{                                 DVMDEMO                                    }
{          Demonstration Program for Digital Voltmeter Routines              }
{ ------------------------------------------------------------------------- }
(* DVMDEMO.PAS 1.0 ©1992 JHJ *)


USES
   Crt,DvmUnit,SerialTk;


CONST
   Holding : boolean = false;
   UsrAbort = 255;
   ESC = #27;


{ main polling loop with simple numerical display of input }

procedure DisplayInput;
var
   DvmData : DvmBlock;
   V : DvmRec;
   Literal : DvmStr;
   CurrDecPt : byte;
   E : integer;
   Ch : char;
begin
 { initialize }
   Ch := #0;
   Holding := false;
   WriteSerial( 0 );                          { set DVM to Run, no dec point }
   FlushSerialBuffer;
   writeln( 'STARTING...' );
 { running loop }
   repeat
   { --- get the next input from the DVM and display it --- }
      if not Holding then
         begin
            GetVal( DvmData,E );                          { get raw input bytes }
            if E <> 0 then
               begin
                  write('GETVAL ERR = ',E);               { catch timeout, etc. }
                  write(' [Esc to quit, any other to continue.] ' );
                  Ch := Upcase(ReadKey);
                  writeln;
               end else
               begin                                    { display input from DVM }
                  ParseDvm( DvmData,V );          { extract info from raw bytes }
                  Literal := DvmValStr( DvmData );    { also make char string }
                  { write results }
                  write('[val ',V.Mag,' dp ',V.DP,']  ');
                  if V.Mark then write(' MARK ');
                  if V.Ovr then write(' OVERFLOW ');
                  write( Literal );
                  writeln;
               end;
         end; {of if-not-holding }
```

Fig. 6-46 Continued.

```
   { --- check if keyboard input --- }
      if KeyPressed then
         begin
            Ch := UpCase(ReadKey);
            { write the keyboard char }
            TextAttr := lightcyan;
            writeln(#13#10,Ch);
            TextAttr := yellow;
            { handle the command }
            case Ch of
               'H' : begin
                        SetDvmHold( true );                    { set DVM to Hold }
                        Holding := true;
                     end;
               'R' : begin
                        SetDvmHold( false );                    { set DVM to Run }
                        Holding := false;
                     end;
               '0'..'5' : SetDvmDP( ord(Ch) - 48 );     { set DVM decimal pt }
               end; {case}
         end; { of if-keypressed }
   until Ch = ESC;
end;


{ ----- setup ----- }

procedure SetComm;
var
   P : byte;
   R : word;
begin
   write( 'Serial Port: 1 for COM1, 2 for COM2: ');
   readln( P );
   if (P=1) or (P=2) then OpenSerial( P ) else Halt(0);
   write( 'Baud rate: ');
   readln( R );
   SetUART( R,Npar,8,1 );                              { in Serial Toolkit unit }
end;

BEGIN
   ClrScr;
   TextAttr := lightgreen;
   writeln( 'DVM TEST   Press a key to begin.  ESC to interrupt and exit.');
   if ReadKey <> ESC then
      begin
         { setup }
            SetComm;
            EnableUART;                                { in Serial Toolkit unit }
         { running }
            TextAttr := yellow;
            DisplayInput;
         { close things down before exiting }
            DisableUART;                               { in Serial Toolkit unit }
            CloseSerial;
      end;
   TextAttr := 7;
   ClrScr;
END.
```

A more substantial example of how to use the DVM and its software is the DvmScope program in chapter 7, a simple but useful data acquisition program.

Summary

This chapter complements the earlier chapters by providing a collection of hardware circuits and modules that can be used to connect a computer to the "real world."

Interface circuits are basic in the sense that they are needed in order to interconnect data acquisition devices and computers. To this end, several parallel interface circuits and a general-purpose serial interface module were presented.

To get the most out of any data acquisition system, a good grasp of principles is essential. Some of the most important theoretical foundations of digital systems were reviewed. Because practically all digital systems are sampled systems, attention has to be given to what goes on in the frequency domain. The importance of controlling the frequency response in D/A and especially in A/D conversion was stressed. A variety of analog filter circuits were presented. Although intended primarily for use with conversion circuits, the filters can also be used for many other applications.

The operating principles of A/D converters were discussed. Two A/D converter modules were presented: a simple but versatile eight-bit converter block, and a 4½-digit digital voltmeter circuit.

Digital-to-analog conversion was looked at, and a general-purpose D/A converter module was presented. The usefulness of the multiplying DAC as a general circuit element—a digitally controlled analog potentiometer, in effect—was pointed out.

The chapter concluded with a fairly extensive example: a complete digital voltmeter system. The DVM consists of several smaller modules that can be combined to create a variety of useful data acquisition instruments, from a simple voltage-monitoring system to a complete DVM with readout. Finally, software to receive data from the DVM and to control it was presented.

The notion throughout this chapter of "building blocks" is meant seriously. Few of the circuits are of much use alone and by themselves, but they can be combined in many ways to build rather sophisticated systems. This is also by far the most versatile and flexible approach. You can tailor instruments to your needs, which are probably not the same as the next fellow's, instead of the other way around as you might have to do with packaged systems. And isn't this kind of flexibility what personal computing is all about?

Setting up systems

This chapter considers some aspects of data acquisition systems. The difference between a system and its component parts lies in its synergy, the "working together" or coordinated interactions among its components. This involves two levels: the infrastructure, so to speak, of interconnections that are the pathways for interaction; and the superstructure that coordinates and runs the overall system, namely the application software.

Earlier chapters looked at a number of ways to implement the infrastructure of data acquisition systems, concentrating on methods that use the standard I/O ports of a PC. This chapter goes to a more basic level and looks at the mundane but crucial matters of interconnections, cabling, and the like. It is a sad fact that these things are often overlooked or done as something of an afterthought, even in some rather expensive commercial systems.

A short and preliminary look will be taken at the system superstructure, the data acquisition software. That is a vast subject about which books could be written—and many have been. The purpose here is not to explore it in any detail but rather to illustrate how the toolkits and the conversion hardware discussed in this book can be combined into a working system. A simple, but useful, data acquisition program will be examined. Finally, there will be some suggestions for creating your own data acquisition and system control programs.

System planning and setup

The first step in planning a data acquisition system is to decide on the transmission method. Sometimes this will be dictated by existing equipment or other constraints. In certain cases, an instrument might use a proprietary interface and perhaps require a special card to be installed in the computer. When the choice is free, the decision is likely to be between more elaborate systems such as the GPIB instrumentation bus mentioned in chapter 3 and simpler systems using the PC's serial or parallel ports. Another possibility that seems likely to be of importance in the future is the SCSI bus, but it has so far been used chiefly for hard drives and tape systems. Because our concern in this book has been with the standard serial and parallel ports, I'll restrict attention to them.

A basic decision is whether to use serial or parallel methods. Both can be used together, of course. Parallel interfaces are fast and can use simple transmitter and receiver hardware. Against this, they require bulky multiconductor shielded cables and are limited to short distances (10 to 12 feet). Serial interfaces use very simple wiring and can transmit over considerable distances, but are slower and require some form of UART and line drivers. Two important advantages of serial systems are the ability to transmit over very long distances with a modem and telephone line and the ability to use a simple current-loop connection that avoids the grounding problems that afflict many systems. There are no hard-and-fast rules for deciding which method to use. It depends on the application. One of the goals of this book is to present what is needed in order to make an informed choice.

In any case it is helpful to know what serial and parallel ports are installed in a computer. The small program PortList in Fig. 7-1 reads the table of serial and parallel port addresses beginning at memory location 400h and lists them, together with their base addresses in hexadecimal notation. More exactly, it lists what ports the BIOS claims are available. For this reason, on a network server it might not list all of the ports. Some network operating systems reserve some or all of the ports for network printers and modify the BIOS table accordingly.

Another element in planning a system is the choice of the data transmission rate or rates. This depends on the application, and there will at least be a minimum rate. This in turn affects the choice

Fig. 7-1 A utility program to find and list the serial and parallel ports in a PC.

```
{ ------------------------------------------------------------------------ }
{                               PortList                                   }
{   Program to list the serial and parallel ports as reported by BIOS.     }
{ ------------------------------------------------------------------------ }
(* PORTLIST.PAS 1.0 *)

TYPE
   string4 = string[4];

VAR
   Portz : array [0..7] of word absolute $40:0;   { BIOS data area port list }
   K,Nr : byte;
   ComBase,LptBase : word;


{ a simple table-driven routine to make a hex literal from a 16-bit binary }

function Word2Hex( N : word ) : string4;
const
   Chars : array [0..15] of char =
        ('0','1','2','3','4','5','6','7','8','9','A','B','C','D','E','F');
var
   S : string4;
   P : byte;
begin
   for P := 4 downto 1 do
      begin
         S[P] := Chars[ N and $F ];
         N := N shr 4;
      end;
   S[0] := #4;                                      { set string length to 4 }
   Word2Hex := S;
end;


BEGIN
   writeln;
   writeln( 'SERIAL AND PARALLEL PORTS' );
   Nr := 1;
   for K := 0 to 3 do                           { loop through BIOS table entries }
      begin
         ComBase := Portz[K];
         LptBase := Portz[K+4];
         write( 'COM',Nr,' ',Word2Hex(ComBase) );
         writeln( '        LPT',Nr,' ',Word2Hex(LptBase) );
         inc(Nr);
      end;
   writeln;
END.
```

of transmission system. Parallel systems generally operate over a range of rates as determined by the processing time of the devices and are paced by data strobes. Serial systems can operate at any rate sufficient to handle the data stream. The baud rate between serial devices must of course match. For simple connections between a PC and an external device I usually use 9600 to 38,400 baud.

System interconnections

The interconnections are a component in every system. This rule is so obvious that it is often overlooked. An otherwise good system might be crippled by improper connections. Never take them for granted. Think of them as circuit elements in the same sense that resistors and capacitors and ICs are circuit elements (which, by the way, is true).

Noise

Digital circuits are less prone to difficulties from noise and interference than analog systems, but they are by no means immune. On the other hand, digital circuits tend to be a potent source of noise.

Figure 7-2 shows the voltages of the first 50 harmonics of a 5-V square wave with 50 percent duty cycle and 1.5 percent rise and fall times. (The data are calculated, using a 1024-point FFT.) Such a waveform would be typical of a 1-MHz HCMOS clock signal. Because the waveform is symmetrical, the even-order harmonics are essentially zero. The zeroth harmonic is the average or dc value, 2.5 V. The fundamental, Fo, is 1.59 V. Note that the harmonic voltages do not fall below 100 mV until 15Fo.

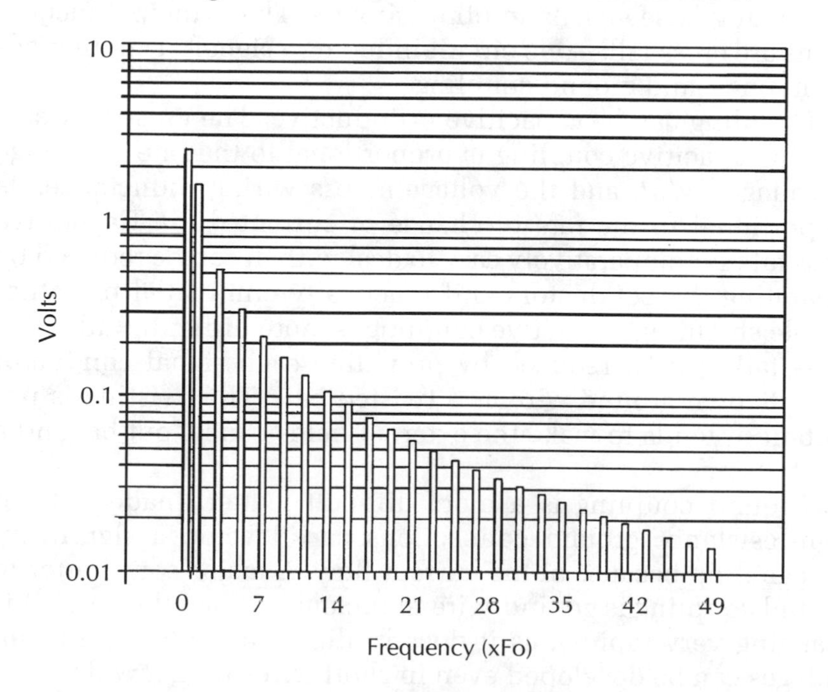

Fig. 7-2 *Frequency spectrum of a typical HCMOS-level square wave.*

Pulses and square waves of very low duty cycles have a $\sin(x)/x$ spectrum. Most of the energy is at frequencies up to $1/\tau$, where τ is the pulse width, but there are significant harmonics up to $3/\tau$. The amplitude of the harmonics is proportional to the average value (the product of the pulse amplitude and the duty cycle).

The significance of these spectra is that they show that there can be a fair amount of energy at frequencies at which practical lengths of interconnecting wiring form fairly good antennas and send some of that energy out into the world. Inadequately shielded and decoupled digital hardware and cabling can create significant amounts of RF interference. A rough but useful test is to listen with a simple AM radio several feet away from the digital system. If noise from the system can be picked up on the radio it is probably too high. Remember that all RF emissions, whether inadvertent or intentional, are subject to FCC regulation. Under the law, correcting harmful interference generated by your equipment is your responsibility and has to be done at your expense.

Within a digital device, problems due to the rate of change of digital waveforms are generally of more significance than the spectrum. (RF energy from one part of a circuit can affect other parts, however.) Rapidly changing signals can cause difficulties by creating false signals in other circuits. The principal mechanisms are cross talk—one circuit inducing a signal in another adjacent one—and ground coupling.

Coupling can be capacitive or inductive. The cross-talk current in capacitive coupling is proportional to the rate of change of voltage dv/dt, and the voltage across wiring inductances is proportional to the rate of change of current di/dt. Capacitive cross talk is comparatively easy to deal with. It can be reduced by separating the conductors and practically eliminated by rather simple shielding. Inductive coupling is more difficult. Inductive cross talk can be reduced by providing each signal conductor with its own ground wire as a twisted pair. The twisting is important; it tends to make the external fields cancel out beyond a short distance.

Ground coupling is a more difficult matter. Inadequate or high-resistance ground connections can develop a significant voltage drop because of the current flow. A major contributor to ground coupling is ground-wire inductance. When the current is changing very rapidly, as it does in digital circuits, significant voltages can be developed even in short wires with low dc resistance. Enough voltage can be coupled through common grounds to disrupt a circuit. The grounds of a circuit should be as short and grouped as closely together as possible. Using a single ground wire for several circuits should be avoided. Wide, flat conductors have less inductance than ordinary round ones, which is part of why ground planes are better than wired grounds.

Inexplicable and unpredictable glitches are often caused by very narrow but very fast spikes, such as a few nanoseconds of output overlap in devices sharing a bus, or a slight skew in the outputs of counters and similar devices. Though brief, the spike current is often quite high—tens of milliamperes or more. If at all possible, use a wideband oscilloscope (100 MHz or more) for troubleshooting.

Grounding and ground loops

Ground noise is by far the greatest problem in digital circuitry. Remember that a ground connection is never truly at zero volts. (Explicitly drawing a resistor on top of each ground line on a schematic diagram can help to visualize the effects.) The key to good circuit layout is to arrange connections so that the grounds of different circuits do not interleave. This confines the inevitable ground noise to a particular circuit and prevents it from becoming part of the input signal to some other part of the circuit. Figure 7-3a shows incorrect grounding in an A/D converter. The ground current from some of the digital circuits passes through the same path as the signal. These currents cause voltage drops in the ground path that are added to the analog input. Figure 7-3b shows the proper method. Analog and digital grounds join at one point only. In critical applications it might be necessary to use separate power supplies to avoid ground coupling through the power supply common ground.

Similar problems can and often do arise in connections between equipment. Most equipment is designed with a grounded mains connection. This can lead to the problem shown in Fig. 7-4. The signal ground, the equipment chassis, and the mains ground form a closed loop that readily picks up noise (especially power-line noise) by transformer action. Beefing up the ground connections often makes the problem worse, because the lower loop resistance permits greater current flow in the loop. A further and not uncommon problem is voltage differences between different mains grounds. Unfortunately, there is no really good cure for this problem. Breaking one or both of the mains ground

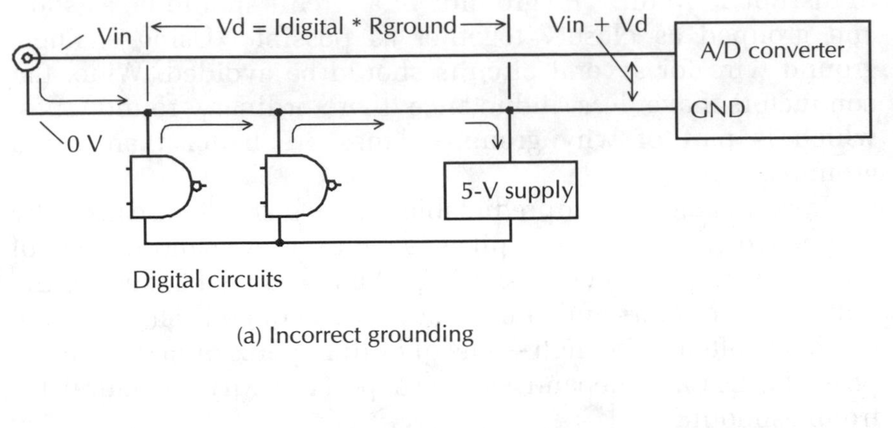

(a) Incorrect grounding

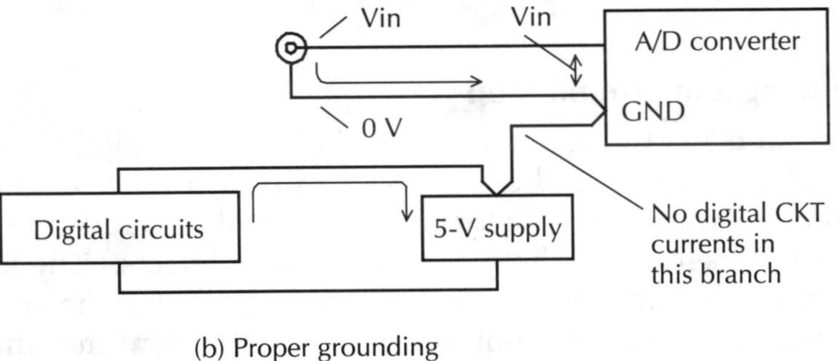

(b) Proper grounding

Fig. 7-3 *Illustration of (a) improper and (b) proper grounding practices.*

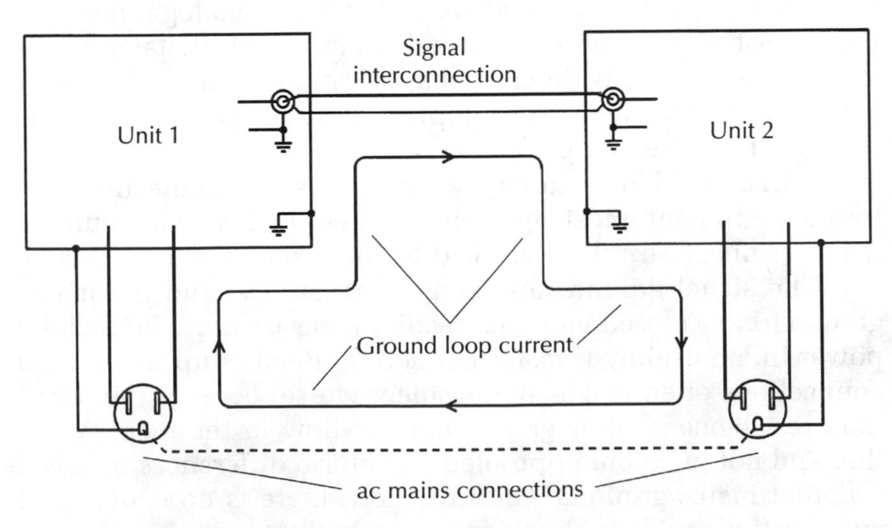

Fig. 7-4 *Example of how ground loops can be formed.*

connections works electrically, but is potentially dangerous and might be illegal.

Ground loops can also exist where several cables connect pieces of equipment together because the multiple ground wires can form closed loops. Keeping the cables as physically close together physically as possible is helpful.

Although ground loops cannot readily be eliminated, they can be made harmless. One way is by means of balanced (differential) signal connections. So far as the signal is concerned, the ground noise is a common-mode voltage and will be ignored if the circuits are properly balanced. The price that is paid is twice as many signal wires and somewhat more complicated line drivers and receivers. Another method is the floating or isolated link. No ground connection at all is used in the signal path. For many decades this has been accomplished through transformer coupling, but optical isolation has become widely used for digital signals.

Ground isolation

Floating circuits have many advantages. Because the ground and signal circuits are separate there are no ground loops. Aside from noise and signal coupling, the avoidance of ground currents is of crucial importance in some small-signal applications such as biological and medical sensors. (Do not fry the patient!) In other cases, particularly industrial applications, the ground references might differ by hundreds or thousands of volts. One way to achieve such isolation is by using an optical rather than an electrical transmission path. Fiber-optic links offer superb performance. As prices fall and the cables become easier to work with they will become the transmission method of choice. Another method is by open light transmission, as in the now ubiquitous remote controls for TV and audio equipment. Strangely, this simple and inexpensive method has not been used very much for data links. A third approach, and one that is especially attractive for the small systems that are the focus of this book, is to use metallic conductors but to include an optical coupler—an LED and a photodetector in one package—in the signal path. Because an LED is a current-operated device, the electrical circuit can be arranged as a current loop, a configuration with many advantages.

The reputation that ordinary optocouplers have for being rather slow is not altogether deserved. The usual implementation of the current loop—a TTL gate with a series resistor as driver and a simple load resistor for the photodetector—needlessly sacrifices performance. Superior results can be obtained with only a little added circuitry.

The detector in an optocoupler is usually a phototransistor in which photons create a base current that is amplified by transistor action. The dynamic response is that of a transistor with current drive. The usual configuration operates the device as a common-collector amplifier (this is true whether the load is in the collector or the emitter lead), and the response suffers considerably due to the Miller effect. This can be avoided by operating the phototransistor with a constant collector-to-base voltage. A good way to do so is with a current mirror as in the receiver circuit in Fig. 7-5b. This circuit switches in less than 2 μs, typically 1.2 to 1.5 μs. The component values shown are for 20-mA loop current.

Ideally the two mirror transistors should be matched for base-to-emitter voltage (V_{BE}) but unselected parts work fairly well. (If you have several transistors on hand, you can select the two with the closest V_{BE} as measured with a diode checker or ohmmeter.) If performance is critical, an inexpensive monolithic transistor array such as the CA3046 is excellent.

Microcomputer current loops are often driven by a TTL device and a series resistor. While inexpensive, it sacrifices performance because the low value of resistance provides poor control of the loop current in relation to cable resistance. A typical 20-mA loop uses a 220-Ω resistor, giving a typical current through the optocoupler of 16 mA. The loop resistance of a 1000-foot cable with 24-gauge conductors is about 51 Ω. This reduces the current by some 20 percent to 13 mA.

The 20-mA driver circuit in Fig. 7-5a uses a current source to overcome this problem. It also ameliorates a second problem. To cause current to begin to flow through the receiver LED the loop voltage must rise from zero to a little over 1 V, and to do this the cable capacitance must be charged. For a 10-nF capacitance (200 to 250 feet of ordinary cable) this is an added delay of half a microsecond in a 20-mA loop, and around 2 μs in a 5-mA loop. But by forcing a very small current through the loop at all times the capacitance is kept charged at the turn-on threshold of the LED. This is the purpose of the precharge resistor R_{PRE}. With 18 kΩ as shown, the off current is around 175 μA. As a bonus, this small precharge current lets you check for a disconnected or broken loop at the transmitter end by measuring the loop voltage because it will rise toward 5 V in that case. You can also test for a shorted cable, because with a short the voltage will be zero or nearly zero.

It is the overall performance that matters. The circuits in Fig. 7-5 were tested with simulated cable resistance up to 68 Ω and capacitance up to 0.1 μF (!), and the worst-case switching time was 2.4 μs—short enough for data at 200 kilobytes per second.

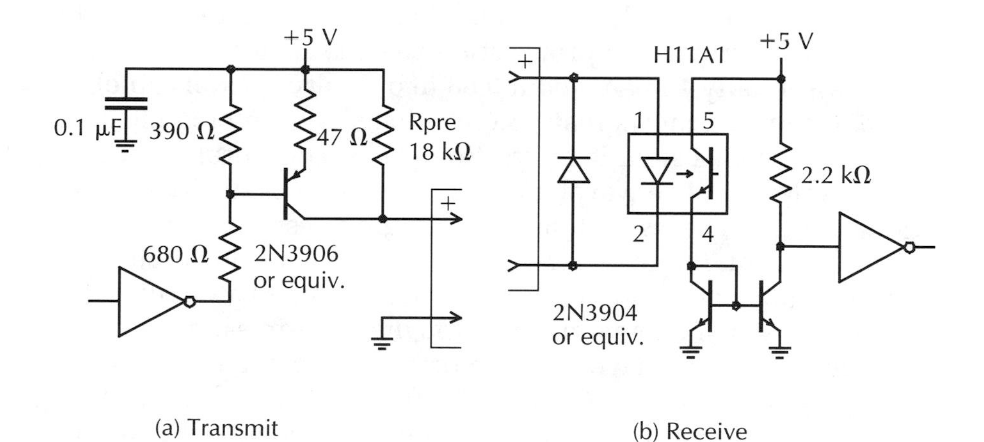

Fig. 7-5 *High-performance 20-mA current loop transmitter and receiver circuits.*

Cabling and line termination

Cables are often thought of as merely the way circuits are connected together. But (to repeat what has been said already) cables are as much a circuit element as resistors and capacitors and ICs are.

The principles of how cables—specifically, transmission lines—behave are well established but, oddly, are often neglected in microcomputer systems. A quick review seems in order. Basically, a transmission line accepts energy at one end and delivers it at the other. Some energy is lost along the way due to the resistance of the conductors, hysteresis losses in the insulation, and radiation. The signal travels at the speed of light, but the speed of light is always less in a conductor than in free space. Moreover, the velocity is often different at different frequencies which can distort the waveshape of a wideband signal. An especially important property of transmission lines is the fact that a portion of the energy might be reflected back toward the source.

It's commonplace, and very useful, to think of electricity as a kind of subtle fluid and circuit elements as discrete, independent entities (lumped constants). But in reality all electric circuits are manipulations of the electromagnetic field. When the size of a circuit is more than a small fraction of the wavelength, the lumped simplifications are inadequate and field phenomena can no longer be neglected. Ordinary wiring is tricky at VHF frequencies and useless for microwaves—waveguides ("plumbing") have to be used instead.

The wavelength of some frequency f in a conductor is c/f, where c is the velocity of light. This can be expressed as $(3 \times 10^8)v/f$, where v is the propagation velocity relative to the vacuum (essentially 1 in air, about 0.68 in ordinary coaxial cable). A little arithmetic shows that practical lengths of interconnecting cables are often a significant fraction (a tenth or more) of the signal wavelength. The wavelength of a 10-MHz signal in a coaxial cable, for example, is 20.4 meters, or slightly less than 67 feet. As a rough rule of thumb, a transmission line can be considered as a lumped constant only if it is shorter than a tenth of a wavelength.

As the signal energy moves through a transmission line, the voltage and current have a certain proportion based on the physical properties of the line. Their ratio is the *characteristic impedance* (Zo), also known as the surge impedance. For a uniform lossless line it is a pure resistance constant with frequency. Real-world cables have some losses and are not precisely uniform; their Zo is slightly reactive and varies somewhat with frequency. Coaxial cables are made to have a specified and stable Zo, commonly 50 to 93 Ω. The Zo of simple twisted pairs is less uniform, and is typically in the neighborhood of 125 Ω.

If the characteristic impedance of a line changes at any point, some of the energy will be reflected back toward the source. (This is true of any electromagnetic transmission medium, actually. The antireflection coating on a camera lens is an impedance-matching network.) This often happens at the ends of the cable. There will be no reflections if, and only if, the impedances of the driving and receiving circuits are equal to Zo. If the mismatch is severe the signal can bounce back and forth through the cable many times. Even moderate mismatches can significantly affect circuit operation, however. Figure 7-6 shows the input impedance of a 10-foot length of 50-Ω coax such as RG-58/U or 174/U for two values of terminating resistance at the receiving end. The upper curve shows a receiver using a single 1-kΩ pull-up resistor, a common practice. Note that at some frequencies the cable loads the driver with just a few ohms—almost with a short. The lower curve shows a better form of ter-

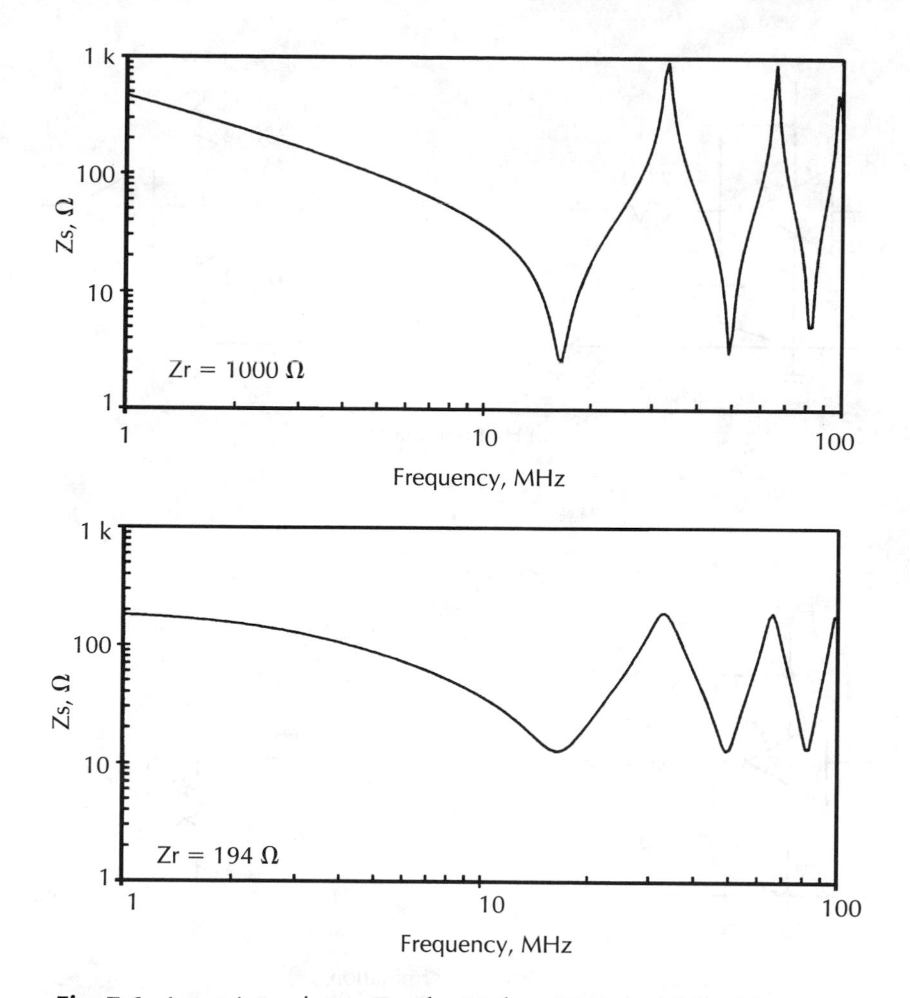

Fig. 7-6 *Input impedance Zs of a 10-foot 50-Ω coaxial cable for two values of terminating impedance Zr.*

mination that will be discussed presently that has an impedance of 194 Ω. The cable input impedance still varies, but much more moderately and smoothly.

Reflections can play havoc with signal timing. Assuming a propagation velocity of 68 percent, the transmission time for a 10-foot length of coax is about 15 ns, and a round-trip reflection takes 30 ns. If there are strong reflections the line might echo for well over 100 ns, which could be devastating to a high-speed data signal.

The moral is: terminate your cables properly! But how? Figure 7-7 shows several methods. The receiving termination is the more critical, and half termination matches only that end. The 50- to

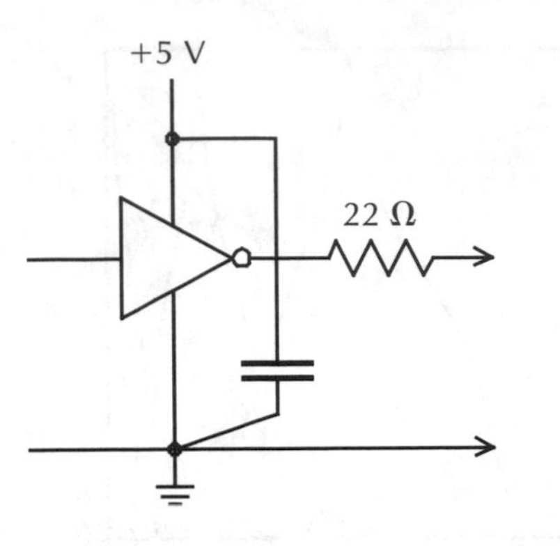

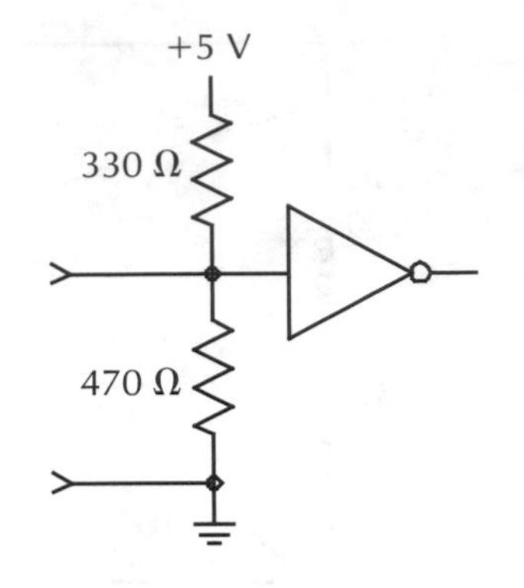

(a) Half termination

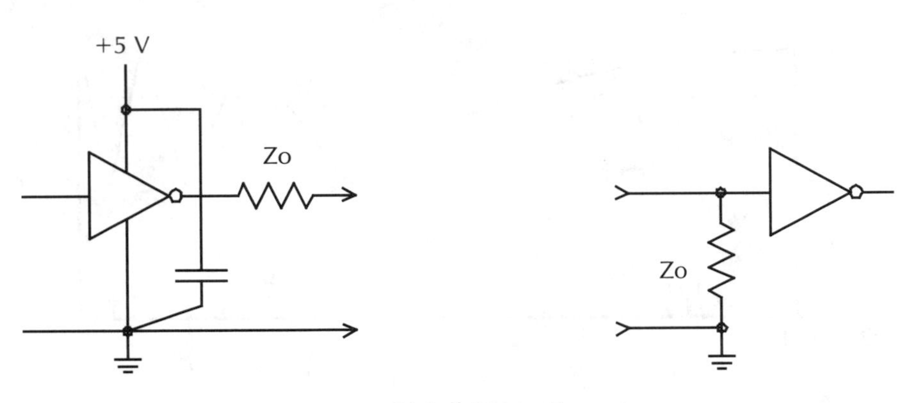

(b) Full termination

Fig. 7-7 *Recommended line termination circuits: (a) half termination for standard TTL and HCT devices; (b) full termination with VMOS driver.*

100-Ω impedance of an exact termination is much too low for ordinary TTL circuits to drive, but a fairly decent compromise is the split termination shown in Fig. 7-7a. The effective termination is 194 Ω. The driving impedance is less important, but still worthy of consideration. Adding a small resistor, say 22 Ω, in series with a TTL driver output significantly improves the driving-point match. In fact, such a series resistor is a good idea for all TTL devices driving more than a foot or so of wire.

For best results, a cable should be matched at both ends. One example of such full termination was shown in the IBM 360 interface in Fig. 6-8. Driver and receiver ICs for this specification are readily available. Another approach is shown in Fig. 7-7b. A high-speed power MOS driver such as the Harris-Intersil ICL7667 drives the line through a resistor approximately equal to Zo. (The output impedance of the 7667 is just a few ohms.) The receiving end is terminated in Zo. The output of the 7667 can swing to nearly +5 V, giving a high signal level at the receiver of 2.5 V which is adequate for LS TTL or HCT CMOS devices. It is not quite enough for the HC or CD CMOS families.

Cabling suggestions

Most of the interconnections in the kinds of data acquisition systems considered here will involve the standard serial and parallel ports. Here are some suggestions for these connections.

Parallel cables should consist of twisted pairs with an overall shield. Standard printer connecting cables generally work well and are readily available. Because the Centronics interface used in the PC parallel port is poorly terminated, there are reflections and ringing on signal transitions. The delay between data line transitions and the data strobe allows time for the lines to settle, but sometimes overshoots and ringing on the strobe line itself cause problems. Parallel interface cables should be kept as short as possible, certainly less than 15 feet.

Serial cables for the RS-232 interface can be quite simple. You can usually get away with most any sort of cable in an RS-232 link, especially on short runs, but good quality cable will help ensure noise-free transmission and minimize interference to other systems. As discussed in chapter 4, only the transmit and receive circuits are necessary for transmission. If you do use some of the RS-232 modem control lines, multiple twisted-pair shielded cable should be used.

The transmission frequencies and the slewing rate of the standard interface confine the signal energy to about 200 kHz even at 38,400 baud. Wave propagation phenomena can be ignored for cable lengths of 300 feet or less at that transmission rate. For lengths up to a few hundred feet the cable behaves much like a lumped capacitance; for longer lengths or thin conductors (less than 24 gauge) the line resistance is also significant.

Cables intended for audio systems are particularly good for RS-232 transmission (which is, after all, mostly in the audio range). They have good shielding and low capacitance. Two-conductor shielded cable such as used for microphones and balanced audio lines works very well, with one conductor for transmit and the other for receive. An even better choice is the twin single-conductor shielded cable widely used for stereo audio systems.

Current loop cabling, as in the example presented earlier, is the least critical of all. The voltage excursions are small, minimizing the effects of line capacitance and radiation. There are substantial current excursions, though, so twisted-pair wiring is strongly recommended in order to minimize the external magnetic field. Two-conductor shielded cable is ideal.

Troubleshooting

"Into every life some rain must fall". However well that describes the general human condition, it seems right on target for computers and systems. Sooner or later—probably sooner—you are apt to find that things are not working right or even not working at all. Here are a few tips that might be of some help in resolving problems.

Computers are exceedingly complicated gadgets and the number of things that can go wrong, and the number of ways that things can go wrong, is immense. This has its serious side as the use of computers to control vital systems continues to expand. As the complexity of a system expands, one of the most powerful tools of quality control, the analysis of possible failure modes, becomes increasingly difficult to apply. The fundamental architecture of ordinary computer hardware and software is combinatorial, which means that the possible states of a system increase extremely rapidly as the size (number of nodes) of the system is increased.

The single most valuable troubleshooting tool at your disposal is the human mind (which, by the way, clearly does not use combinatorial logic, but apparently combinations of parallel processing and priority and "fuzzy" logic). Start by formulating a plan of attack, not by trying a few random things to see if they help. Pay close attention to all the symptoms, remembering that some might be consequences of others. Once you have satisfied yourself that you have grasped what is not working right (and also what is), use these things as clues to the identity of the culprit and play the detective. Take a moment to consider what things might cause these kinds of effects. Deducing possible causes is greatly helped if you have a grasp of the principles involved, which is one reason they have been stressed in the course of this book.

The way to approach ferreting out the source or sources of

problems depends on whether you are wrestling with a new setup and trying to make it work, or are faced with a system that had been functioning properly but no longer does. In the former case, a good method is to divide and conquer. Isolate the components of a system and check them individually. Try using a different PC if one is available. When the problem has newly developed in a working system, on the other hand, the first thing to suspect is some component failure. Another common cause is modifications to the PC or the system, even ones that would not seem to be at all related to the current difficulty. Has any new software or new versions of existing software been installed on the computer? (This applies even to software that seems totally unrelated to data acquisition such as a spreadsheet or a desktop publishing program.) Have any adapter boards been added or changed? If so, be suspicious and investigate.

A first step should always be to make sure that the computer itself seems to be functioning properly—for example, check that the parallel port will operate a printer and the serial port will run a modem. If you are dealing with a new machine, discuss it with the vendor. You might want to review the material on I/O addresses and IRQ lines in chapter 2 for help in detecting port or interrupt conflicts. If all else fails, try removing and reinserting the adapter cards in the PC very carefully. The finger connectors used in the PC bus are not as dependable as true plug-and-socket types.

Don't forget the basics. Make sure that interconnecting cables and connectors are sound. This is the first thing to check if a system begins to behave erratically. If possible, substitute known good cables and see if the difficulties clear up.

Serial problems

The electrical characteristics and connections for RS-232 serial systems were discussed in chapter 3. You might want to look back at them.

Line problems can be checked with an oscilloscope or, to a lesser extent, a voltmeter. The proper levels are shown in Fig. 3-15. There are commercial line-testing devices which have LEDs to display circuit states that might be helpful. Use an oscilloscope if at all possible; it will reveal far more of what is going on. With a little practice you can easily learn to "read" the start, data, and stop bit cells.

A telecommunications program or the terminal emulator

demonstration program in chapter 4 is useful for testing. If a second computer is available it can be set up as a test driver.

The most common serial problem is transmission parameter mismatch. This should be the first thing to check if data gets through but is garbled. Double-check the baud rate, number of data bits, parity, and number of stop bits. Incorrect parity is a particularly common difficulty. The **CommError** function in the Serial Toolkit can help identify which transmission parameters are incorrect. The actual baud rate can be measured with an oscilloscope set to trigger on negative-going signals which will pick out the start bit.

Don't overlook the possibility that an external device expects the modem control lines to be used. If so, be sure that the connecting cables contain those circuits. It's a good idea to check the actual pins used by the external device for the control circuits. Having read chapters 3 and 4 you now know what the control lines are supposed to do and which pins are what, but the designer of the device might not have known this!

Parallel problems

Because of the large number of conductors and the construction of the standard connector, the first step in diagnosing an ailing parallel system should be to check the cables—preferably by substitution with known good ones.

If the parallel port is being used for general I/O and not as a standard Centronics interface, double-check the setup and connections. Don't forget that the control register (PCR) must be set so that the line drivers are in the high state as discussed earlier. If the parallel port is being used for general-purpose interrupt-driven input, check the input strobe (the *ACK line) pulse width. As a test, try lengthening it to, say, 200 µs which should trigger the interrupt on most any computer.

Unreliable or erratic operation despite known good cables and connectors is often caused by noise and ringing. Use a good wideband oscilloscope to check. This is especially likely on fast computers with later versions of the parallel port that accept narrow *ACK strobes—substantial ringing might be taken to be two or more separate strobes.

It's also a good idea to check the actual levels on all of the lines. For reliable operation they must conform to TTL noise margin limits with the external device connected. Measurements at the port connector on the PC without the effects of the loading of an external device will catch only major failures. The low level should be less than 0.8 V, and the high level should be at least 2.0 V. If the voltages are outside these limits, check further. The port drivers might be defective, or the flaw might be in the input circuits in the external device.

Software woes

If the problem seems to be in the software, try another program if possible—for instance, a telecommunications program in the case of the serial port. If that works but code you have written doesn't, go back over your code carefully. If you are using one of the toolkits in this book there should be no problem with the port I/O itself (unless you have modified the toolkit). If the problem arises with some other program, even a commercial program, but things operate correctly with software you have written, congratulate yourself! (Yes, this does happen.)

If you are using interrupts, review the material in chapter 4 carefully, especially regarding the interrupt controller (PIC). Make sure that the interrupt mask is being set correctly, and that the proper EOI is issued at the end of the ISR.

If you are running under Microsoft Windows, try exiting and running the software from the DOS command line. The material in this book should work properly under Windows as a standard DOS program. (If your program has no PIF file, Windows is supposed to assume it might use the BIOS calls, the serial and parallel ports, and address video memory directly.) Remember that the routines in this book assume real-mode operation. Many protected-mode operating systems and DOS extenders claim that running a program as a DOS session is or acts just like real-mode, but glitches are by no means unknown.

Data acquisition software

All data acquisition systems require some sort of overall software to run and manage the system and to gather and process results. This is the place where this book leaves off and some other book would begin.

To summarize and bring together the many topics discussed in this book, let's now put together a very modest but functional data acquisition program. This program, DvmScope, provides an oscilloscope-like graphical display. It is intended for use with the digital voltmeter described in chapter 6. However, Dvm-

Scope also has built-in "test signals" so it can be run by itself for demonstration without any external input. Figure 7-8 is a screen dump of the program running in test mode, showing two sine waves at two different *Y*-axis scales.

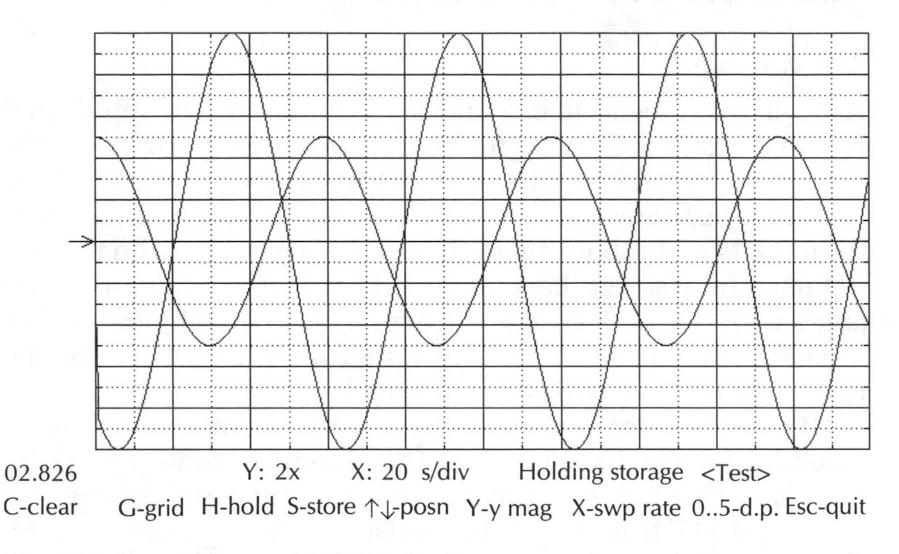

Fig. 7-8 *Screen dump of the DVMScope program in stand-alone test mode.*

When used with the DVM presented in chapter 6 it turns a PC into a digital storage oscilloscope. The combination isn't fast, but it does have very high resolution. It is ideal for monitoring processes or quantities (such as temperature or pressure) over minutes or hours.

DvmScope draws on lessons learned in the course of a number of more elaborate systems I have developed for various applications. The basic display format has evolved somewhat over several years and over many megabytes of data. It has proved to be quite useful and is offered as an example for your consideration.

The main purpose of the program, though, is for illustration and also for your study, modification, and extension. You could revise it to operate with other input devices such as a digitizing tablet or some instrument you might have. The simple eight-bit A/D converter module in chapter 6 could be used instead of the DVM, allowing up to several hundred points per second. The program can also be extended to allow input data to be stored in memory or on disk for later use. Other possibilities include adding other output capabilities, such as a plotter. You might also find the graphical routines useful for displaying various kinds of computed numerical data sets.

Program commands

DvmScope has a few simple commands, listed in Table 7-1. At start-up the options are D to use the DVM as the input source, and T to use the built-in test signals. Choice T would be used to run the program by itself for demonstration.

Table 7-1 Commands for the demonstration program.

Command	Function	
Start-up commands		
D	Use DVM input	
T	Test mode (dummy data; use no external input)	
	R Create random data	
	S Create sine wave data	
Esc	Exit to sys	
Running commands		
C	Clear screen	
G	Turn grid on, off	
H	Hold (also sets DVM to hold)	
S	Turn storage on, off (off = clear screen after each sweep)	
Y	Step through *y* "gain" values, 1-2-5-10-20-50	
X	Step through *x* "sweep rate" values, 1-2-4-10-20	
Esc	Quit display, go to start-up menu	
Cursor keys:		
	Up arrow	Shift baseline (*y* position) up
	Down arrow	Shift baseline down
	Home	Set baseline to 0 (midscreen)

The display can be run in either of two modes. In STORAGE ON mode, curves are written continuously and accumulate on the screen. STORAGE OFF clears the screen after each screenful. You can toggle back and forth between the two modes by pressing the S key. Pressing C at any time clears the screen.

When running on an EGA or VGA video system, curves are displayed against a graph-paper-like grid, which has proved to be very useful. The grid can be turned on or off at anytime by pressing G.

The H key toggles HOLD on and off (that is, "hold" or "run").

In HOLD mode, data are not written to the screen. If the DVM is being used, it too is set to hold the last measured value.

The up and down cursor ("arrow") keys can be used to move the curve baseline up or down, like the *Y*-position control on an oscilloscope. PgUp and PgDn move in larger ten-pixel increments, and Home resets the baseline to midscreen.

The *Y* (vertical) and *X* (horizontal or time) scale factors can be controlled. A 1-2-5 sequence is provided, like an oscilloscope. Pressing Y steps to the next higher scaling or "gain," and pressing X does likewise for the horizontal "gain."

The program

The DvmScope program is listed in Fig. 7-9. It uses the DvmUnit module developed in chapter 6 and the serial toolkit module developed in chapter 4. (The parallel toolkit module from chapter 5 could be substituted.) It also draws on two Turbo Pascal units, Crt and Graph. The latter contains the Borland graphics interface (BGI) system introduced with Turbo Pascal 4 and included with Turbo C 1.5 and later.

The program itself is divided into several functional sections. In a more elaborate version these would be separate modules. The *graphics display* routines set up the basic display frame and screen prompts, and accept data values for plotting on the screen. The *input data handling* routines accept data from the DVM with the help of the DvmUnit and the serial toolkit modules and prepare it for use by the graphics routines. The *interactive user interface* is the basic dispatcher and controller for the program.

The display routines are divided into initialization, frame (axes, grid, etc.), and data drawing sections. The basic frame is drawn by the **Axes** routine. The colors were chosen with some care, taking into account the perceptual characteristics of human vision. Higher contrast and less fatigue results when the background of light-emitting displays is black (the opposite is true for things viewed by reflected light, such as a printed page, where dark images on a white or light background is better). The acuity (sharpness) of vision depends somewhat on color and is greatest for green, yellow, and white, so these are used for the curves and other prominent elements. These colors also seem to be slightly forward on a two-dimensional display. Blue, on the other hand, seems to recede a little so it is used for the grid. The color palette in EGA and VGA video systems is used to turn the grid on and off

Fig. 7-9 *DVMScope, a sample data acquisition and display program.*

```
{ ---------------------------------------------------------------------------- }
{                              DVM SCOPE                                       }
{      Sample Data Acquisition Program using the Digital Voltmeter            }
{ ---------------------------------------------------------------------------- }
(* DSCOPE.PAS 1.1 ©1992 JHJ *)

USES
   Crt,Graph,SerialTk,DvmUnit;

{$V- }

TYPE
   float = real;

CONST
{ control char equates }
   CR = #13;
   ESC = #27;
   SPACE = #32;
{ screen layout }
   Xspan = 600;
   Xorg = 40;
   StatusLine : byte = 24;                          { location of status line }
{ colors }
   PlotColor : byte = yellow;                                        { data }
   AxesColor : byte = green;                                         { axes }
   GridColor = blue;                                                 { grid }
   LegendColor : byte = lightgreen;                         { scale legends }
   OvlyColor = white;                                       { marker, etc. }
   PromptColor = yellow;                                    { status, etc. }
   HelpColor = 7;                                           { help messages }
{ initialized state variables }
   ShowGrid : boolean = true;
   Store : boolean = true;
   Holding : boolean = false;
   CommOpen : boolean = false;
{ "gain" multipliers }
   Ymult : array [1..6] of byte = (1,2,5,10,20,50);
   Xmult : array [1..5] of byte = (1,2,4,10,20);

VAR
   GrafDriver,GrafMode,GrafErr : integer;
   Ymid,Yspan,Ymajor,Yminor : word;                         { plot constants }
   Yscale : float;
   Ypos : word;                         { Y axis position (baseline offset) }
   Ymag : byte;                                          { Y magnification }
   Xstep : byte;                                             { X increment }
   Xmag : byte;
   DvmMode : boolean;      { false = use dummy test data; true = use DVM data }
   TestDelay : word;        { optional delay to simulate DVM measurement rate }
{ ----- graphics init and setup ----- }

{ initialize screen layout variables according to video type }

procedure SetExtents;
begin
   if GetMaxY > 350 then
      begin                                                  { VGA 640x480 }
```

Fig. 7-9 ***Continued.***

```
          Yspan := 440;
          Ymid := 220;
          StatusLine := 29;                  { VGA 480 has 30 8x16 text lines }
          WindMax := 7903;                   { reset limits in Turbo CRT unit }
       end
   else if GetMaxY > 200 then
       begin                                 { EGA 640x350, Hercules (348 lns) }
          Yspan := 320;
          Ymid := 160;
       end
   else begin                                                    { CGA 640x200 }
          Yspan := 180;
          Ymid := 90;
       end;
   Ymajor := Yspan div 10;
   Yminor := Yspan div 20;
end;

procedure SetColors;
begin
   if GetMaxColor < 15 then
       begin
          PlotColor := 1;
          AxesColor := 1;
       end;
end;

procedure SetYscale( Gain : word );
begin
   Yscale := (Ymid/20000.0) * Gain;
end;

procedure InitScaling;                                { set scaling factors }
begin
   Ypos := Ymid;
   Ymag := 1;
   SetYscale( 1 );
   Xmag := 1;
   Xstep := 1;
end;

procedure ActivateGraf;                           { go from text to graphics }
begin
   DirectVideo := false;
   SetGraphMode( GrafMode );
   if ShowGrid then SetPalette( 1,GridColor ) else SetPalette( 1,0 );
end;

procedure ActivateText;                           { go from graphics to text }
begin
   ClearDevice;
   RestoreCrtMode;
   DirectVideo := true;
   TextAttr := 7;  ClrScr;
   Delay(100);                                       { settling time for crt }
end;

function InstallGraf : integer;            { initialize BGI graphics system }
begin
```

Fig. 7-9 Continued.

```
   GrafDriver := 0;                                       { driver 0 = autodetect }
   InitGraph(GrafDriver,GrafMode,'');
   InstallGraf := GraphResult;
   if GrafDriver = CGA then GrafMode := CGAhi;            { if CGA, force 640x200 }
   SetColors;
   SetExtents;
   InitScaling;
   ActivateText;
end;


{ --------- screen layout and formatting --------- }

procedure DrawFrame;
begin
   SetColor( AxesColor );
   SetLineStyle( SolidLn,0,NormWidth );
   Rectangle( 40,0,639,Yspan );
end;

procedure DrawGrid;
var
   K,X,Y : word;
begin
   if GrafDriver = CGA then Exit;       { remove this to allow grid with CGA }
   SetColor( GridColor );
   X := 60;
   Y := Ymajor;
   for K := 1 to 9 do                                     { major divisions }
      begin
         Line( 41,Y,638,Y );                                   { horizontal }
         Line( (40+X),1,(40+X),Yspan-1);                         { vertical }
         inc( X,60 );
         inc( Y,Ymajor );
      end;
   SetLineStyle( UserBitLn,$8888,NormWidth );          { sparsely dotted line }
   X := 30;
   Y := Yminor;
   for K := 1 to 10 do                                    { minor divisions }
      begin
         Line( 41,Y,638,Y );
         Line( (40+X),1,(40+X),Yspan-1 );
         inc( X,60 );
         inc( Y,Ymajor );
      end;
   SetLineStyle( SolidLn,0,NormWidth );
end;

procedure DrawTicks;
var
   K,X,Y : word;
begin
   SetColor( AxesColor );
   X := 60;
   Y := Ymajor;
   for K := 1 to 9 do                                     { major divisions }
      begin
         Line( 41,Y,47,Y );                                          { left }
```

Fig. 7-9 Continued.

```
         Line( 633,Y,639,Y );                                  { right }
         Line( (40+X),1,(40+X),5 );                            { top }
         Line( (40+X),Yspan-5,(40+X),Yspan-1 );                { bottom }
         inc( X,60 );
         inc( Y,Ymajor );
      end;
   X := 30;
   Y := Yminor;
   for K := 1 to 10 do                                { minor divisions }
      begin
         Line( 41,Y,45,Y );
         Line( 635,Y,639,Y );
         Line( (40+X),1,(40+X),3);
         Line( (40+X),Yspan-3,(40+X),Yspan-1 );
         inc( X,60 );
         inc( Y,Ymajor );
      end;
end;

procedure MarkBaseline( Visible : boolean );    { draw or "undraw" an arrow }
var
   C : byte;
begin
   C := GetColor;
   if Visible then SetColor( AxesColor ) else SetColor( 0 );
   Line( 20,Ypos,39,Ypos );
   Line( 30,Ypos+4,39,Ypos );
   Line( 30,Ypos-4,39,Ypos );
   SetColor( C );
end;

procedure Axes;
begin
   DrawFrame;
   DrawGrid;
   DrawTicks;
   MarkBaseline( true );
end;

procedure ToggleGrid;
begin
   ShowGrid := not ShowGrid;
   if ShowGrid then SetPalette(1,GridColor) else SetPalette(1,0);
end;

procedure DrawMarker;                 { indicate when event marker triggered }
var
   C,X : word;
   L : LineSettingsType; { declared in BGI }
begin
   C := GetColor;
   X := GetX;
   GetLineSettings( L );
   SetColor( OvlyColor );
   SetLineStyle( DottedLn,0,NormWidth );
   Line( X,10,X,Yspan-10 );
   SetLineStyle( L.LineStyle,L.Pattern,L.Thickness );
   SetColor( C );
```

Fig. 7-9 Continued.

```
end;


{ ----- screen prompts ----- }

procedure WritePrompt;
begin
   GotoXY( 1,StatusLine+1 );
   TextAttr := HelpColor;
   write( 'C-clear G-grid H-hold S-store  ' );
   write( #24#25'-posn  Y-y mag  X-swp rate  0..5-d.p. Esc-quit' );
   TextAttr := PromptColor;
end;

procedure ClearStatus;
var
   P : byte;
begin
   GotoXY( 1,StatusLine );
   for P := 1 to 79 do write( SPACE );
   write( CR );
end;

procedure WriteStatus;
begin
   TextAttr := LegendColor;
   GotoXY( 21,StatusLine );
   write( 'Y: ',Ymult[Ymag]:2,'x    ');
   write( 'X: ',(20 div Xmult[Xmag]):2,' s/div    ' );
   TextAttr := PromptColor;
   if Holding then write( 'HOLDING  ' ) else write( '          ' );
   if Store then write( 'STORAGE  ' ) else write( '          ' );
   if not DvmMode then write( '<Test>' );
end;

procedure Readout( S : DvmStr );         { running report of current reading }
begin
   GotoXY( 1,StatusLine ); write( S,' ' );
end;

function VerifyNoInput : char;                          { handle timeout error }
begin
   ClearStatus;
   write('NO INPUT [Esc to quit, any other to continue.]' );
   VerifyNoInput := Upcase(ReadKey);
   ClearStatus;
   WriteStatus;
end;

{ ----- screen drawing ----- }

procedure Redraw;
var
   X,Y : word;
begin
   X := GetX;
   Y := GetY;
   ClearDevice;
   Axes;
   WritePrompt;
```

Fig. 7-9 Continued.

```
   WriteStatus;
   SetColor( PlotColor );
   MoveTo( X,Y );
end;

procedure DrawValue( Value : integer );
var
   X,Y : integer;
begin
   X := GetX;
   inc( X,Xstep );
   if X > 639 then
      begin
         if not Store then Redraw;
         MoveTo( Xorg,GetY );
         X := Xorg + Xstep;
      end;
   Y := Ypos - round(Yscale * Value );
   if Y < 0 then Y := 0;
   if Y > Yspan then Y := Yspan;
   LineTo( X,Y );
end;

procedure SetBaseline;
begin
   MarkBaseline( false );                        { erase current baseline marker }
   case ReadKey of
      #71 : Ypos := Ymid;                           { home resets to midscreen }
      #72 : dec( Ypos );                            { up }
      #73 : dec( Ypos,10 );                         { pg up }
      #80 : inc( Ypos );                            { down }
      #81 : inc( Ypos,10 );                         { pg dn }
      end; {case}
   if Ypos < 10 then Ypos := 10;
   if Ypos > Yspan-10 then Ypos := Yspan-10;
   MarkBaseline( true );                                   { draw new marker }
end;


{ ----- get and dispatch input events ----- }

function GetFirstValue : boolean;                               { true = ok }
var
   DvmData : DvmBlock;
   V : DvmRec;
   E : integer;
begin
   if DvmMode then GetVal( DvmData,E ) else TestGetVal( DvmData,E );
   if (E <> 0)  then GetFirstValue := false else
      begin
         ParseDvm( DvmData,V );
         MoveTo( Xorg,Ypos-round(Yscale * V.Mag) );  { cursor to first point }
         GetFirstValue := true;
      end;
end;
```

Fig. 7-9 Continued.

```
procedure DisplayInput;
var
   DvmData : DvmBlock;
   V : DvmRec;
   E : integer;
   Ch : char;
   Done : boolean;
begin
{ initialize }
   Ch := #0;
   Done := false;
   Holding := false;
   InitScaling;
   Redraw;
{ get a first value }
   if not GetFirstValue then Done := (VerifyNoInput = ESC);
{ running loop }
   while not Done do
      begin
      { --- get the next input from the DVM and display it --- }
         if not Holding then
            begin
               if DvmMode then GetVal( DvmData,E ) else
                  begin
                     TestGetVal( DvmData,E );
                     if TestDelay > 0 then Delay( TestDelay );
                  end;
               if E <> 0 then Done := (VerifyNoInput = ESC) else
                  begin                                { display input from DVM }
                     ParseDvm( DvmData,V );
                     DrawValue( V.Mag );
                     if V.Mark then DrawMarker;
                     Readout( DvmValStr(DvmData) );
                  end;
            end; {of if-not-holding}
      { --- check if keyboard input --- }
         if KeyPressed then
            begin
               Ch := UpCase(ReadKey);
               { handle user commands }
               case Ch of
                   #0 : SetBaseline;
                  ESC : Done := true;
                  '0'..'5' : if DvmMode then SetDvmDP( ord(Ch) - 48 );
                  'C' : Redraw;
                  'G' : ToggleGrid;
                  'H' : begin
                           Holding := not Holding;
                           if DvmMode then SetDvmHold( Holding );
                        end;
                  'S' : Store := not Store;
                  'X' : begin
                           inc( Xmag );
                           if Xmag > 5 then Xmag := 1;
                           Xstep := Xmult[Xmag];
                        end;
                  'Y' : begin
                           inc( Ymag );
```

Fig. 7-9 Continued.

```
                                      if Ymag > 6 then Ymag := 1;
                                      SetYscale( Ymult[Ymag] );
                                   end;
                             end; {case}
                          WriteStatus;
                       end; {of if-keypressed}
                 end; { of while-not-Done }
end; { of DisplayInput }


{ ----- setup ----- }

procedure SetUpComm;
var
   P : byte;
   R : word;
begin
   repeat
      write( 'Serial Port: 1 for COM1, 2 for COM2: ');
      readln( P );
   until (P=1) or (P=2);
   OpenSerial( P );
   CommOpen := true;
   write( 'Baud rate: ');
   readln( R );
   SetUART( R,Npar,8,1 );                               { in Serial Toolkit unit }
   EnableUART;
end;

procedure SetUpDvm;
begin
   if not CommOpen then SetUpComm;
   WriteSerial( 0 );                              { set DVM to Run, no dec point }
   FlushSerialBuffer;
end;

procedure SetUpTest;
var
   Ch : char;
begin
   write( 'Test signal:  S-sine  R-random ' );
   Ch := Upcase( ReadKey );
   writeln( Ch );
   if Ch = 'S' then TestType := tmSine;
   if Ch = 'R' then TestType := tmRand;
   write( 'Sample delay, ms: ' );
   readln( TestDelay );
end;

function MainDispatch : boolean;                               { false = exit }
var
   Ch : char;
begin
   ClrScr;
   writeln( 'DVM-SCOPE DEMONSTRATION' );
   writeln;
   writeln( 'D-digital voltmeter    T-test mode     Esc-exit' );
   writeln;
```

Fig. 7-9 Continued.

```
      Ch := Upcase( ReadKey );
      if Ch = ESC then MainDispatch := false else
         begin
            writeln(Ch);
            writeln;
            DvmMode := (Ch = 'D');
            if DvmMode then SetUpDvm else SetUpTest;
            MainDispatch := true;
         end;
end;

BEGIN
   if InstallGraf <> 0 then
      begin
         writeln('Error installing graph driver.');
         Halt(0);
      end;
   while MainDispatch do
      begin
         ActivateGraf;
         DisplayInput;
         ActivateText;
      end;
   if CommOpen then
      begin
         DisableUART;
         CloseSerial;
      end;
END.
```

(in **ToggleGrid**) by setting color 1 to blue (on) or black (off). Because CGA cards have somewhat marginal resolution and do not support full color, the grid is suppressed if such a card has been detected.

A single routine, **DrawValue**, handles both drawing and scaling—the input routines simply pass a data value to it. Global variables hold the current scaling factors, and are set by **SetExtents** in accordance with the graphics hardware in use. The data scaling is based on data values of $\pm 20,000$ full scale as provided by the DVM; this value, held as an immediate operand in **SetYscale**, can be changed for other devices. The actual pixel coordinates are determined by the global scaling, the data value, and a "gain" multiplier. Values are drawn as line segments between the current value and the immediately preceding one in order to create a continuous curve.

There is a certain degree of conflict in data acquisition and display programs between the graphics and various prompts and other text. There is no single best solution for all cases. In this program, the choice is to maximize the graphical area. Its purpose is, after all, to display the data.

Screen prompts are done through the BIOS for speed and simplicity, even though the BGI system has a variety of fixed and scalable fonts. This is handled by the Turbo Crt unit. Ordinarily it writes text character codes directly to video memory, but because it is running in graphics mode the BIOS must be used. Setting the Crt variable **DirectVideo** to FALSE does this. The text clipping limits in *WindMax* are also changed for VGA operation because the Crt unit default is 25 text rows and VGA at 640 by 480 resolution has the equivalent of 30 text rows.

Input data handling from the DVM and interactive input from the user is essentially the same as in the DvmDemo sample program in chapter 6. The **DisplayInput** subroutine uses a simple polling loop to get user input and to receive data. Data values are obtained by calling routines in the DVM unit, either **GetVal** or, for artificial "test" data, **TestGetVal.** DVM unit routines are likewise used for DVM control functions such as HOLD/RUN and decimal point placement. Extensions to this sample program could add calls to subroutines to save data values in memory or write them to disk. If memory storage is implemented, a useful addition would be the ability to "play back" previous screens for review.

Two points about **DisplayInput** are worth noting. Data values are drawn as line segments between the current point and the previous one, but of course the very first data value has no predecessor. **GetFirstValue** handles this situation. It moves the graphics cursor to the first point but does not draw anything. The second thing to be noted is provision for bailing out if the data input is absent or interrupted. **VerifyNoInput** lets the user choose to quit or to try again (after fixing a loose connector, for instance). The actual time-out supervision is handled in the DVM unit.

A main dispatching loop and a few start-up routines complete the program. The main program body (the function **main()** in C programs) consists of a check of BGI initialization, the dispatching loop, and routines for an orderly exit.

Developing data acquisition programs

DvmScope is offered as both a model and a framework. As a framework, you could modify it and extend it to use other input devices, accumulate data in memory, provide disk I/O, and so forth. As a model, it shows some techniques you can draw on for data acquisition and analysis software of your own design.

There is one consideration which, though present in nearly any program, is of particular importance in data acquisition software. That is the matter of partitioning the program. This is a lot like creating code modules, but partitioning is expressed in the structure of the overall program design rather than in explicit modules as such. Notice that the DvmScope program is divided into several related functional areas, marked in the program listing (Fig. 7-9) by section headings. In a simple program like DvmScope the choice of partitions is fairly obvious. Graphics screen functions clearly constitute one group, for example. If data were being written to and from disk, those would form yet another. It is not always easy to decide exactly how a program should be partitioned, and in some cases routines might be so closely related that the partitioning is not as readily seen or as easy to effect. The propinquity of data input and user input code in the DisplayInput subroutine is an example of this.

There are some other considerations that are very important in data acquisition software but usually not in other kinds of programs. These have to do with the quantity and especially the rate of data input. This is not a problem in the case of DvmDemo and the DVM with its three readings per second, but in many actual data acquisition applications it is a major issue—and sometimes a tough one to solve. The toolkits in this book were written with this in mind, and that is why they waste as little time as possible. There are also a number of techniques that can be used in programs for dealing with these issues, but a discussion of them lies beyond the scope of this book.

Summary and prospect

Data acquisition systems, like all systems, depend not on their component parts alone, but also on the interconnections between them and on the overall organization and control of the system. We have looked at a number of issues at both of these levels of system function.

It is trite but true that nothing is stronger than its weakest link. All too often, system interconnections are shortchanged in the planning and setup of systems. Some of the problems that can arise from improper interconnections and various ways to avoid them were discussed. In a more positive vein, suggestions and recommendations were offered for the best ways to hook up the kinds of equipment that I have dealt with in this book.

The best transmission setup is of little value without suitable software to run the system and to carry out data presentation and analysis. A small sample program illustrated some of the ways in which the hardware and software of the earlier parts of the book could be put to work. Although a thorough treatment of the software side of data acquisition is beyond the scope of this book, some preliminary suggestions were offered.

With that, our look at small but effective data acquisition systems comes to a close. Now it is, in a sense, your turn. It is hoped that what has been said here will offer some inspiration and a few helpful tips you can use to create your own systems. I have not the slightest doubt that we have only begun to scratch the surface of what small computer systems can do. The principles are by now pretty well explored. Moving ahead depends on bright ideas. If anything in this book leads you to have some, I will feel well repaid.

Glossary

address In computers, a number that specifies or enables connection of a bus to a particular location in memory or I/O space.

address bus Transmission path for addresses in a computer.

aliasing Corruption and distortion of a signal in a sampled system due to an inadequate sampling rate.

alignment Used in electrical filter theory to designate a particular form of response.

analog A representation of one configuration by another; specifically, some quantity by a continuous variable. Often used as an antonym for digital.

ANSI American National Standards Institute (formerly ASA, American Standards Association). Clearinghouse for standards and participates on behalf of United States in international standards activities.

assembler Customary name for compilers of assembly language, which is basically the command set of a CPU expressed by symbols (names).

balanced In analog and digital transmission systems, two signals of opposite polarity symmetrical with respect to ground or common. Balanced circuits provide superior noise rejection.

baud From J. M. E. Baudot, inventor of the first practical teletypewriter. The reciprocal of the duration of the smallest interval in a serial message transmission. Each such interval might contain more than one bit of information.

BCD Binary Coded Decimal. Four-bit binary representation of the decimal digits 0 to 9.

block device An I/O channel that handles bytes in bursts or groups (blocks).

CCITT [French for:] International Consultative Committee for Telephone and Telegraphy. An organization for international standards for telephone and similar networks.

character device An I/O channel that handles bytes singly.

comparator An amplifier whose output switches between two states depending on whether the input is greater than or less than some reference value.

CRC Cyclic Redundancy Check. A powerful polynomial checksum technique based on the uniqueness of certain prime factors. Widely used for error detection.

current mirror Two or more transistors with base terminals connected together. An input transistor operates with 100 percent feedback (base tied to collector). Current into this transistor develops a base voltage that in turn causes a similar current flow in the remaining transistor(s).

data bus Transmission path for information in a computer. In conventional (von Neumann) computer architecture, program instructions and program data are intermixed and both use the same data bus.

DCE Data Communication Equipment. In RS-232 serial systems, a modem. *See also* DTE.

digital 1. A discrete representation of a quantity by numerical values. 2. Generally, systems using only two states, usually implementing Boolean algebra. *See also* analog.

DTE Data Terminal Equipment. Devices that use an RS-232 serial link but are not DCEs; for example, computers, printers, etc.

EIA Electronic Industries Association. Electronic manufacturers' trade association.

EMI Electromagnetic Interference. The emission or pickup of noise energy by a circuit or device. If at the mains frequency, it is usually called *hum*.

far In the Intel 80x86 processor, a full address (twenty to thirty-two bits depending on the processor). Contrasted with *near*.

FDX *See* full-duplex.

frame A unit or block of information in a transmission system, typically composed of data, synchronizing information, and in some cases identification and/or error-detection information.

full-duplex A transmission system in which two parties can transmit simultaneously.

ground loop A closed path, common to one one or more signal circuits, into which noise can be induced.

group delay The rate of change of phase as a function of frequency in a sinusoidal signal passing through a network. The phase is defined as between the output and the input signal.

half-duplex A transmission system in which two parties take turns sending.

handshake A means of controlling exchange of information among devices; it often incorporates acknowledgment of successful or failed reception.

HDX See half-duplex.

high-level language Programming languages that use constructs more or less similar to natural or mathematical language, from which detailed machine language instructions are created by a compiler. Some examples are FORTRAN, BASIC, COBOL, Pascal, C, APL, Modula-2, and Ada.

IEEE Institute of Electrical and Electronic Engineers. A professional engineering association.

interrupt Method of changing program flow in response to some event outside of the currently executing instructions.

interrupt service routine A subroutine intended to be invoked by an interrupt rather than by an ordinary call. Typically it carries out whatever function the interrupt is being used for.

IRE Institute of Radio Engineers. Now the middle "E" in IEEE, a merger of the IRE and the Institute of Electrical Engineers.

ISO International Standards Organization. Members are national standards groups; for example, ANSI.

line coding Form of the signal in a transmission system; for example, on-off, positive-negative (as in RS-232), and so on.

module In programming languages, code that can be combined with other modules to form a complete program. Some languages (for example, Modula-2 and Turbo Pascal) enforce strict interface requirements to insulate the code in modules from one another to minimize unforeseen and unintended side effects when modules are combined.

near In the Intel 80*x*86 processor, the least significant sixteen bits of an address. Contrasted with *far*.

Nyquist limit From H. Nyquist, who demonstrated that in sampled systems the input signal must be sampled at a rate at least twice the highest frequency component of signal to avoid aliasing.

object module Machine (binary) language together with information that allows it to be combined or merged (linked) with other object modules to form a complete machine-language program. Assemblers ordinarily emit object modules rather than complete programs.

operational amplifier An amplifier, often (but not necessarily) with a differential input, operated with such a large amount of negative feedback that the overall circuit behavior is determined primarily by the feedback network.

parallel Simultaneous transmission of more than one unit of information; typically, eight or sixteen bits at a time.

parity The addition of one or more bits to force a group of data bits to yield an even or odd sum. A simple but weak method of error detection.

reconstruction In digital signal processing, the process of recovering an analog signal from its digital representation.

record In high-level programming languages, a declaration of several variables that can be referenced as a single entity or unit.

recursion Generally, self-reference; specifically, a subroutine calling another instance of itself. *See also* recursion.

reentrancy Ability of a subroutine to be called repeatedly without affecting results of other calls of it.

RFI Radio Frequency Interference. High-frequency EMI.

RS-232 A standard for the interconnections between modems and other devices using serial transmission. *See also* DCE and DTE.

serial Units of information (usually bits) transmitted sequentially in time. Used loosely to refer to TTY frame coding and RS-232 transmission.

static variable A variable used by a subroutine that retains its value between calls. Contrasted with temporary local variables which are abandoned at the conclusion of a subroutine.

structure *See* record.

symbol Something that stands for something else. (1) In programming, a name or "handle," often mnemonic, for a machine-language operation code or address. (2) In communications, a character.

TTY Teletypewriter. Also used for start-stop synchronizing bit frame coding.

unbalanced In analog and digital transmission systems, a circuit composed of one signal line plus ground or common.

vector In computers, an address. An interrupt vector is a variable holding the (entry) address of the interrupt service routine.

Bibliography

Caristi, Anthony J. 1989. *IEEE-488 general purpose instrumentation bus manual.* San Diego: Academic Press.

Eggebrecht, Lewis C. 1990. *Interfacing to the IBM personal computer.* 2nd ed. Indianapolis: Howard W. Sams.

Fletcher, John G. 1982. An arithmetic checksum for serial transmissions. *IEEE Transactions on Communications* COM-30:247.

Glass, L. Brett. 1989. Inside EISA. *BYTE* 14(November):417.

Graeme, Jerald G., Gene E. Tobey, and Lawrence P. Huelsman. 1971. *Operational amplifiers: design and applications.* New York: McGraw-Hill.

Morse, Stephen P. 1982. *The 8086/8088 primer.* 2nd ed. Hasbrouck Heights, NJ: Hayden Book Co.

Nelson, Mark R. 1992. File verification using CRC. *Dr. Dobb's Journal* 17(5):64.

Peterson, W. W., and D. T. Brown. 1961. Cyclic codes for error detection. *Institute of Radio Engineers Proceedings* 49:228.

Peterson, W. Wesley. 1972. *Error-correcting codes.* 2nd ed. Cambridge, Mass.: MIT Press.

Proakis, John G., and Dimitris G. Manolakis. 1988. *Introduction to digital signal processing.* New York: Macmillan.

Rose, Charles G. 1990. *Programmer's guide to Netware.* New York: McGraw-Hill.

Sheingold, Daniel H., ed. 1974. *Nonlinear circuits handbook.* Norwood, Mass.: Analog Devices, Inc.

Sheingold, Daniel H., ed. 1980. *Transducer interfacing handbook.* Norwood, Mass.: Analog Devices, Inc.

Index

Software Offer

If you're interested in running any of the programs shown in this book, you might want to order the companion disk. This disk contains executables, source code, object modules, and handy "readme" files for all the code shown in this book, and is available in double density 3½" or 5¼" media, for IBM-compatible computers, for a price of $24.95, plus $2.50 shipping and handling. A Pascal compiler is needed if you want to make modifications to the source code.

YES, I'm interested. Please send me:

____ copies 5¼" disk (#8705436), $24.95 each . . . . $ ________

____ copies 3½" disk (#8705444), $24.95 each . . . . $ ________

____ TAB/McGraw-Hill computer catalog (free with purchase) . . . . $ FREE

Shipping & Handling: $2.50 per disk in U.S./Canada $ ________

Please add applicable state and local sales tax. $ ________

TOTAL $ ________

❑ Check or money order enclosed made payable to TAB/McGraw-Hill

Charge my ❑ VISA ❑ MasterCard ❑ American Express ❑ Discover Card

Acct No. ______________________________ Exp. Date ________

Signature ______________________________

Name ______________________________

Address ______________________________

City ______________________ State ______ Zip ________

TOLL-FREE ORDERING: ☎ 1-800-822-8158

(outside the U.S. and Canada call 1-717-794-2191)

or write to TAB/McGraw-Hill, Blue Ridge Summit, PA, 17294-0840

Prices subject to change. Orders outside the U.S. and Canada must be prepaid in U.S. dollars and include an additional $5.00 for postage and handling.

TAB-4355